The Problem of Consciousness

New Essays in Phenomenological Philosophy of Mind

The Problem of Consciousness

New Essays in Phenomenological Philosophy of Mind

Edited by
Evan Thompson

University of Calgary Press
Calgary, Alberta, Canada

ISSN 0229-7051 ISBN 0-919491-29-4

University of Calgary Press
2500 University Drive NW
Calgary, Alberta
Canada T2N 1N4
www.uofcpress.com

Library and Archives Canada Cataloguing in Publication

The problem of consciousness : new essays in phenomenological philosophy of mind / edited by Evan Thompson.

(Canadian journal of philosophy. Supplementary volume ; 29)
Includes bibliographical references and index.
ISBN 0-919491-29-4

1. Consciousness. 2. Phenomenology. 3. Cognitive science.
I. Thompson, Evan, 1962- II. Series.

B105.C477P76 2004 126 C2004-905276-4

We acknowledge the financial support of the Government of Canada through the Book Publishing Industry Development Program (BPIDP), the Alberta Foundation for the Arts and the Alberta Lottery Fund—Community Initiatives Program for our publishing activities.

Printed and bound in Canada by Houghton Boston.
This book is printed on acid-free paper.

Cover design by Mieka West. Typesetting by Samuel Smith Esseh.

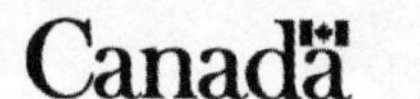

Canada Council for the Arts
Conseil des Arts du Canada

Table of Contents

CANADIAN JOURNAL OF PHILOSOPHY
Supplementary Volume 29

Introduction

EVAN THOMPSON

Phenomenology, like analytic philosophy, is not a philosophical position, but one of the dominant philosophical traditions to arise in the twentieth century. Inaugurated by Edmund Husserl (1859–1938), this tradition includes major twentieth-century European philosophers (such as Martin Heidegger, Maurice Merleau-Ponty, and Jean-Paul Sartre), as well as important North American and Asian exponents. In recent years, a new current of phenomenological philosophy has emerged in Europe and North America, one which goes back to the source of phenomenology in Husserl's thought, but addresses issues in contemporary analytic philosophy of mind, philosophy of psychology, and cognitive science.[1] The articles in this volume belong to this current.

Much of the writing in this new current of phenomenological philosophy of mind takes its point of departure from the so-called 'explanatory gap' for consciousness, also known as the 'hard problem' of consciousness.[2] It is not too difficult to state in general terms what the explanatory gap is supposed to be, but the moment one tries to formulate it more precisely one enters the thickets of philosophical

---

1 Representative examples of this trend are the recent collection of articles entitled *Naturalizing Phenomenology: Issues in Contemporary Phenomenology and Cognitive Science* (Petitot et al. 1999) and the new journal *Phenomenology and the Cognitive Sciences*.

2 See Roy et al. (1999) for an account of the explanatory gap and its relationship to phenomenological philosophy of mind. Roy summarizes this account in his contribution to the present volume.

controversy. In general terms, the problem is that there seems to be a gap between the bio-behavioural and abstract, functional characteristics of cognition and the subjective and experiential aspects of mental processes. This sense of a gap is evident in the fact that there is still no well-developed scientific account of consciousness and its relation to brain activity (or to bio-behavioural processes more generally), despite considerable scientific advances in understanding the 'mind-brain' and growing scientific respectability for the topic of consciousness. Beyond this well-recognized situation, however, there is little agreement on what exactly the explanatory gap is – whether it is epistemological or metaphysical, for instance, or derived from misleading ways of thinking about consciousness.[3]

It is important to observe that there is no unified position on these issues within phenomenological philosophy of mind; on the contrary, there is considerable disagreement and debate. Nevertheless, phenomenological philosophers of mind do agree on one salient point, which is that discussions of the explanatory gap in philosophy, as well as models of consciousness in cognitive science, have often been impoverished in their conceptual analyses and descriptive accounts of the phenomenology of conscious experience. Thus one underlying conviction of phenomenological philosophy of mind is that much more precise, phenomenological accounts of the subjective-experiential phenomena themselves are needed, if significant progress is to be made on the explanatory gap.

Each of the essays in this volume is animated by this concern. The first two, by Jean-Michel Roy and Denis Fisette, address epistemological issues associated with the concept of phenomenological description. Roy's topic is the apparent challenge to phenomenology posed by the problem of the so-called 'myth of the given' (the myth of cognitive or experiential content free from any interpretative activity); Fisette's is the conceptual status and character of phenomenological description in relation to empirical psychology and the problem of consciousness as discussed in analytic philosophy of mind. The third essay, by Dan Zahavi, examines the relationship between intentionality and

3 See the articles collected in Block et al. (1997).

experience and through this examination criticizes David Chalmers' distinction between the easy problems and the 'hard problem' of consciousness. The next two essays, one by Shaun Gallagher and the late Francisco Varela, the other by Robert Hanna and Evan Thompson, explore issues at the convergence of phenomenology and cognitive neuroscience. Using time-consciousness as an example, Gallagher and Varela show that phenomenology and cognitive neuroscience can integrate their research in a productive manner beneficial to each field. Hanna and Thompson, drawing on some of the same neuroscientific research discussed by Gallagher and Varela, investigate the 'spontaneity' of consciousness from a theoretical perspective known as 'neurophenomenology.' Finally, Depraz and Cosmelli present an extended analysis of empathy and intersubjectivity, in which they develop a multifaceted second-person theoretical perspective on the explanatory gap or hard problem of consciousness, beyond the dichotomous first-person / third-person framework in which this problem is usually formulated.

References

Block, N., O. Flanagan, and G. Güzeldere, eds. (1997). *The Nature of Consciousness: Philosophical Debates* Cambridge, MA: MIT Press.

Petitot, J. P., F. J. Varela, B. Pachoud, and J.-M. Roy, eds. (1999). *Naturalizing Phenomenology: Issues in Contemporary Phenomenology and Cognitive Science*. Stanford: Stanford University Press.

Roy, J.-M., J. Petitot, B. Pachoud, and F. J. Varela (1999). "Beyond the Gap: An Introduction to Naturalizing Phenomenology." In J. Peitot, F. J. Varela, B. Pachoud, and J.-M. Roy, eds., *Naturalizing Phenomenology: Issues in Contemporary Phenomenology and Cognitive Science*, 1–80. Stanford, CA: Stanford University Press.

CANADIAN JOURNAL OF PHILOSOPHY
Supplementary Volume 29

Phenomenological Claims and the Myth of the Given

JEAN-MICHEL ROY

1. The Cognitive Relevance of Husserlian Phenomenology

Over the past twenty years, Husserlian phenomenology has increasingly drawn the attention of the cognitive community, thereby leading to the emergence of what might be called a phenomenological trend within contemporary cognitive studies. What this phenomenological trend really amounts to is however a matter of debate. The reason is that it embodies, in fact, a multifaceted reflection about the relevance of Husserlian phenomenology to the current efforts towards a scientific theory of cognition, and, to a lesser degree, about the reciprocal relevance of these efforts to the fate of the Husserlian tradition. Indeed, it covers a wide array of perspectives on these questions, ranging from tentative demonstrations of the cognitively misleading character of Husserl's ideas, inasmuch as they would incarnate the same foundational errors as cognitivism (Hall and Dreyfus 1982), to diametrically opposed views arguing that the naturalist bent of contemporary cognitive science is ill-conceived, and that only the brand of transcendentalism defended by Husserl can provide it with adequate foundations (cf., for instance, Villela-Petit 1999). In a recent attempt (Roy 2003a) to survey this emerging but already complex area of cognitive research, I proposed to see it as divided around three crucial points:

- the specific type of Husserlian phenomenology being vindicated (Husserl's, or that of any one of his numerous and variously critical followers);

- the negative or positive nature of the cognitive relevance advocated for it;[1]
- the content of the claim of cognitive relevance being made: the essential division options, here, are between locating that relevance in the substance of Husserlian accounts of cognitive phenomena or in their epistemological form (especially in the idea of so-called first-person science), or in both.

1.1. The Claim for a Phenomenological Level of Investigation

One of the most innovative ways of arguing for the positive relevance of both the substance and the epistemological form of Husserlian accounts of cognition is to advocate the introduction of a specific level of phenomenological investigation of a Husserlian kind into the naturalistic framework of contemporary cognitive science. This leaves nevertheless quite a number of options open, depending on the type of Husserlian phenomenology taken into consideration, on the interpretation offered for it, and on the number and nature of the elements of such an interpretation which are effectively incorporated into cognitive science. Any plea in favour of the cognitive relevance of Husserlian phenomenology should demonstrate, if not a high degree of precision on these points, at least a high degree of awareness of the necessity for being precise about them, for fear of building a moot case. Articulating a defence of the cognitive relevance of Husserlian phenomenology is accordingly a quite involved undertaking, and it is arguable that none of the recent proposals to this effect yet qualifies as such. All of them remain at the moment essentially programmatic, offering at best partial attempts to

1 Dreyfus's claim that the critique of Husserl's representationalism by Heidegger lays bare the foundational errors of cognitivist representationalism is an example both of the negative relevance to cognitive science of Husserl's phenomenology and of the positive one of Heidegger's phenomenology, each of which is considered as falling under the general category of Husserlian phenomenology (understood as the sort of phenomenological inquiry originating in Husserl's work).

illustrate the fruitfulness of the research program they promote. And this is precisely the reason why the phenomenological trend in cognitive science is more adequately viewed as an emerging direction of research, than as a recent turn taken by a segment of the cognitive community.

One of these programmatic claims for the introduction of a phenomenological level of investigation of a Husserlian kind into contemporary cognitive science was outlined in "Beyond the Gap: An Introduction to Naturalizing Phenomenology" (Roy et al. 1999) in connection with a number of empirical, theoretical and critical investigations carried out by its authors, as well as other researchers working along similar lines.[2] On the basis of further inquiry (Roy 2000, 2000/02), this outline can be shortly restated in a more personalized way as the conjunction of the following theses:[3]

1. In order to fully account for its object, an adequate theory of cognition has to explain the existence of cognitive phenomenological properties, understood as the way things look to a cognitive system in virtue of its cognitive activity; in other words, as what there is *for* a cognitive system *from* its own perspective (versus both what there is independently of the cognitive system, and what there is for a system independently of its own way of apprehending it).

2. These phenomenological properties have to be explained on the basis of other types of properties of various degrees of abstractness, but including ultimately chemico-physical ones. This requirement is what makes the approach faithful to the naturalist commitment of contemporary cognitive science.

---

2 Including the editor of this special issue of the *Canadian Journal of Philosophy*; see Petitot et al. (1999).

3 For lack of a better name, I have elsewhere (Roy 2000) referred to this specific way of stating the claim for the introduction of a Husserlian level of phenomenological investigation as the RPPV hypothesis, a designation simply derived from the initials of the authors of the 1999 essay where it is first laid out.

3. Phenomenological properties so understood are essentially subjective, in the sense of being by very definition relative to specific (and, ultimately, individual) cognitive systems. They thus cannot be shared: a phenomenological property of a cognitive system S1 cannot also be that of a cognitive system S2, although nothing in principle prevents the two cognitive systems from having phenomenological properties which are in many respects similar to each other.

4. Phenomenological properties can be divided into different categories, and especially between internal and external, as well as conscious and unconscious ones.

5. Conscious phenomenological properties are the ones a system is aware of having: unless blind, we are for instance aware of a visual manifestation of our environment. Through such awareness, a cognitive system is endowed with a distinctive knowledge of its phenomenological properties. What makes this knowledge distinctive is, at least (leaving aside the question of the possible specificity of its nature and truth-value), that it can no more be shared than those properties themselves, because it can only be obtained by their subject. Nobody else than me can have my awareness of how roses smell to me, even though many others may probably have an awareness of it very similar to the one I have. Some phenomenological properties are thus seen as coming with what is traditionally labelled an "epistemic privilege," although in the limited sense just stated: it is in particular not assumed that the knowledge provided by consciousness is of a higher quality than any other, or that it cannot be approximated by other forms of knowledge.

6. Prior to being explained in naturalist terms, phenomenological properties have to be established with detail and accuracy. Scientific explanation requires a rigorous *explanandum*. The specific claim being made in this respect is that phenomenological *explananda* cannot be adequately obtained without using reports emanating from the subject of the

phenomenological properties at stake. Such reports are what deserve *proprio sensu* the title of phenomenological description. Accordingly, a phenomenological description is not only a description of some phenomenological properties (as defined above), but a description carried out by the subject of these phenomenological properties *and* based on the specific knowledge attached to its subject status. As such, it has therefore two essential epistemological features. It is a purely descriptive kind of knowledge, anxious to carefully avoid the introduction of any feature not consciously apprehended as well as any explanatory move; its only goal is to determine what is consciously apprehended faithfully and without trying to explain it. And it is also a first-person knowledge, in the fundamental sense that the subject of the investigation is identical with the subject of the investigated properties. Recapitulating the various specifications introduced so far, a (cognitive) phenomenological description is therefore defined as a knowledge: 1) of what there is *for* a cognitive system *S*, 2) *from* the perspective of *S*, and 3) *as reported* by *S* (and consequently based on the consciousness *S* has of what there is for it and according to it). This restricted use of the notion of phenomenological description is in accordance with the history of the notion of phenomenology, and especially with the Husserlian tradition, although it has certainly been, and can legitimately be, used in a more liberal way to designate any study of phenomenological properties.[4]

7. The reason for granting such an indispensable role to the subject in the establishment of phenomenological *explananda* lies precisely in the epistemic privilege it is endowed with, namely the fact that it benefits from that knowledge of phenomenological properties traditionally called consciousness.

4 I think, however, that using the word for designating the phenomenological properties themselves – a widespread habit in the cognitive literature – is strongly inadequate.

This reason is in fact twofold.

To the extent that a cognitive theory must account for phenomenological properties in general, the knowledge provided by consciousness must naturally be taken advantage of, unless it can be shown systematically unreliable or useless. To the extent, for instance, that we want to know what there is *for* a cognitive system *S* and *according* to its own perspective when it sees a red apple, and that it is endowed with a specific knowledge of this visual phenomenon by virtue of being conscious of it, this knowledge is to be naturally considered as an important source of information, unless it can be proved to be either fundamentally unreliable or useless. Being fundamentally unreliable makes it necessarily useless. But it could be useless without being utterly unreliable if scientists could appeal to another source yielding on average better results, such as purely behavioural data. Relying on the knowledge of its visual apprehension of red apples that *S* obtains through consciousness in no way implies that this knowledge cannot be equivocated, and that its visual phenomenological properties are necessarily as it apprehends them consciously. The possibility that consciousness might be in this specific case overly distorting or partial, and that the true nature of its phenomenological properties remains essentially beyond its reach is not ruled out. The cognitive scientist is here like a detective who happens to have a number of witnesses at his disposal and decides, in order to elucidate the real nature of the criminal events he investigates, to make use of their testimonies on the basis of the reasonable assumption that none of them is utterly handicapped or biased, even though some of them may very well turn out to be substantially equivocated.

However, to the extent that a cognitive theory must specifically account for *conscious* phenomenological properties, for phenomenological properties *as* their subject knows them through consciousness, the need to use that subject's reports, if available, is a much more commanding one. Indeed, phenomenological properties *as they are known* through consciousness are no longer simply a source of information in order to reach the true nature of phenomenological properties *as they are,* but the object itself of the inquiry. The scientist is no longer like a detective interested in the true nature of criminal events, but like a reporter interested in conveying to his newspaper audience what it was like to see those horrible events, whatever they really were. If conscious knowledge is indeed unique, in the second case there can be no better way in principle to reach the goal of cognitive theory than obtaining from *S* himself an expression of its knowledge. Indeed, by virtue of this uniqueness assumption, any knowledge obtained from a different source is necessarily different. Accordingly, any systematic exclusion of phenomenological reports must either be the result of breaking with such assumption or express resignation to a stopgap solution. A natural way of refusing the assumption of uniqueness is to accept a strong notion of empathy, according to which the cognitive scientist can share to a large extent the conscious phenomenon under study.[5] And a standard stopgap solution consists in trying to reconstruct in a purely indirect way (using behavioural and neurobiological data) what *S* is conscious of, taking advantage of the generic cognitive similarities that the cognitive theorist might share with it, as well of the resources of his imagination. A drawing or computerized image, produced by this scientist, of what *S* is visually conscious of when *S* looks at an apple is an instance of such a reconstruction and is to be contrasted with a drawing produced by *S* itself. Although only an approximation by definition, this picture provides the scientist with a knowledge similar in kind with that of *S*, and one that can perhaps be refined to the point of making the factor of approximation negligible. Finally, it is arguable that the reason for renouncing phenomenological descriptions in favour of purely

5 It is arguable, though, that admittance of empathy only amounts to weakening the uniqueness assumption, and not to rejecting it.

reconstructive ones cannot have to do with the imperfections of conscious knowledge itself, but only of its reportability, a point often misstated in standard criticisms of introspection. Indeed, if what we are concerned with is what *S* is conscious of, it does not make sense to dispute the reliability of *S*'s consciousness, but only *S*'s capacity to describe it faithfully.

8. In light of these considerations, it appears that the claim for the introduction of a phenomenological level of description is necessarily related to the three following assumptions:

   a) the knowledge provided by consciousness is not a systematically unreliable or useless source of information for characterizing the nature of phenomenological properties;
   b) humans do not have a capacity for empathy significant enough to serve as an adequate basis for the characterization of specifically conscious phenomenological properties;
   c) the reportability of the sort of knowledge obtained through consciousness is not sufficiently imperfect to recommend the systematic exclusion of phenomenological descriptive reports from the process of establishing phenomenological properties.

   It is important to underline that none of the opposite claims is nevertheless assumed to be true. In other words, the phenomenological claim under consideration is committed neither to seeing conscious knowledge as fully reliable or its reportability as unproblematic, nor to denying the possibility of a certain degree of empathy in humans.

9. Even though the definition of a phenomenological description it is associated with is in essential respects faithful to the Husserlian tradition, it is insufficient to justify seeing this phenomenological claim as a specifically Husserlian one. Indeed, not only is such definition also associated with squarely

anti-Husserlian principles, such as the commitment to naturalism previously mentioned, but it is also not specifically Husserlian per se. Other forms of phenomenological investigations did and could subscribe to it. Accordingly, the reason for vindicating a special relation with Husserlian phenomenology lies in the content of the phenomenological descriptions themselves. As a matter of fact, the core of the hypothesis, in this respect, is that certain aspects of the cognitive phenomenological descriptions provided by the Husserlian tradition are essentially correct, and that elaborating on these descriptive principles will lead to a characterization of phenomenological properties that make them much more easily amenable to a naturalist explanation along the lines of contemporary cognitive science. The stakes of the hypothesis are therefore quite high: no less than bridging the so-called explanatory gap. And its basic *motto* is: get adequate phenomenological descriptions by going Husserlian, and phenomenological properties will prove to be far less resistant to integration into the naturalist framework of contemporary cognitive science than they look.

1.2. Founding the Claim

The hypothesis just summarized is a foundational one, in the sense that it makes claims about what some of the fundamental tenets of a scientific theory of phenomenological properties should be. As such, it needs complementary developments of two different kinds. On the one hand, these fundamental tenets must be articulated with appropriate details and reflection so as to make each one of them as thoroughly clarified and objection-free as possible, a task that requires rather extensive critical work. On the other hand, it must be applied to specific cases and shown to be fruitful and successful in these contexts.

It is important to emphasize the necessity for both kinds of development, as some cognitive theorists seem inclined to overlook the insufficiency of empirical fruitfulness and success for accepting a scientific hypothesis: indeed, the history of science has clearly taught us that competing hypotheses can be equally satisfactory from an empirical point of view, but unequally so from a foundational one. In addition, the critical aspect of the task of articulating the content of this specific foundational hypothesis calls for special attention. For most of the elements entering in its formulation have been the target of severe criticism in the course of twentieth-century philosophy, as well as of epistemological reflection at large. Moreover, these criticisms, ranging from the rejection of the very existence of phenomenological properties to the denial of the possibility of providing a strictly descriptive characterization of them, have played a major role in central areas of the analytic tradition and have in fact represented pivotal elements of its purported antagonism with the phenomenological one. Consequently, unless it provides specific and solid rebuttals of these criticisms, any claim for the cognitive relevance of Husserlian phenomenology is subject to serious charges of philosophical naivety and obsolescence, even though it might invoke convincing demonstrations on the empirical side. Indeed, as long as those principled objections have not been taken care of, the significance of such empirical results remains disputable on various counts. This is the reason why the original version of the phenomenological claim under consideration was cautious to make provisional remarks about the need for critical considerations (cf. Roy et al. 1999, sec. 6), although it did not pretend to offer any. And my ambition in this paper is precisely to contribute to the fulfilment of this critical task, thereby pushing one step further an attempt to develop this original version along more personal lines initiated in other essays (cf. Roy 2000, 2000/02).

1.3. The Problem of the Myth of the Given

Among the various objections to be confronted, one has undisputable priority: is Husserlian phenomenology guilty of subscribing to the "myth of the given" repeatedly denounced over more than a half-century of post-Husserlian philosophy by a variety of thinkers as diverse in intellectual orientation and temper as Wittgenstein, Neurath, Sellars, Kuhn, or Derrida? Indeed, as indicated by this far from exhaustive enumeration, the idea that givenness is an illusion, a typical example of the sort of artefacts resulting from misconceived philosophical and epistemological thinking, has been advanced on so many sides that it is no longer possible to assume that there is such a thing as the given without providing a serious rebuttal of these criticisms. There are in addition special reasons why cognitive scientists favouring the integration of a level of phenomenological investigation of a Husserlian kind should feel concerned by this critical requirement. On the one hand, the very notion of consciousness seems to be intrinsically related to the idea of givenness, inasmuch as consciousness seems to have been always more or less implicitly assimilated to a kind of direct and immediate cognitive relation, including in the Husserlian tradition. On the other hand, this particular tradition, and Husserl especially, often makes an explicit and crucial use of the idea of givenness through a commitment to one form or another of intuitionism. Thus most crusaders against the myth of the given take it for granted that Husserlian phenomenology succumbs to its illusions. A good illustration of this standard attitude is provided by a critical remark from Dennett, who rejects the appeal to a Husserl-style phenomenological investigation in the study of phenomenal consciousness partly on the ground that "like so many attempts to strip away interpretation and reveal the basic facts of consciousness to rigorous observation, such as the Impressionist movement in the arts and the Introspectionist psychologies of Wundt, Titchener, and others, Phenomenology has failed to find a single, settled method that everyone could agree upon" (Dennett 1991, 44). Indeed, such rejection is based on the idea that Husserl's phenomenology presupposes the existence of a given construed as a cognitive content free from any interpretive activity, and consequently as something apprehended in a purely passive way.

The issue is therefore straightforward, and one of utmost importance for the phenomenological claim summarized above: if indeed the given is mythical, and if indeed Husserlian phenomenology hinges on it, cognitive science should definitely better stay away from Husserlian phenomenology. And the strategy to be followed for solving it is no less straightforward. It is first necessary to clarify the myth of the given, so as to disentangle its various aspects as well as its numerous versions. One should then examine, focusing in priority on Husserl's original formulation of phenomenology, whether any of these aspects and versions corresponds to the notion of the given Husserlian phenomenology is relying on, or whether Husserlian phenomenology fails to succumb to the myth of the given, either by devising a different notion, or, appearances notwithstanding, by proving to pay only lip service to the idea of the given. If it does succumb to the myth, the next step is to determine whether the specific phenomenological claim defended here inherits this error, or whether it is preserved from it though its rejection of a number of Husserlian principles. And here again, the crucial aspect of the problem is in relation to Husserl's original version of phenomenology. Finally, if the phenomenological claim does inherit the problem, three additional tasks have to be addressed:

a) assessing the soundness of the claim that the given is mythical: It might very well be that the denunciation of the myth of the given under consideration is based on unacceptable arguments.
b) assessing the vulnerability of the phenomenological claim: Even if it is supported by acceptable arguments, it might very well be that the problem can be circumscribed, and in particular that Husserlian phenomenology provides interesting resources for doing so.
c) assessing the damage to the phenomenological claim: Even if no remedy can be found, it might also very well happen, finally, that the problem turns out to be insufficiently central to jeopardize the essence of the phenomenological claim.

To the best of my knowledge, nobody has so far undertaken this systematic inquiry. It clearly extends beyond the limits of this contribution, however. Consequently, I will limit my ambition to investigate a few aspects of such a research program.

2. Circumscribing the Myth of the Given

2.1. A Fluctuent Criticism

What exactly is the myth of the given? Two questions are involved here: What is the given? And: What is mythical about it? The fact is that there is no real consensus on these two points. The expression 'myth of the given' is an umbrella one covering a large number of criticisms similarly guided by the idea that the notion of givenness is flawed, but diverging in various degrees as to what is flawed, in what respects, and for what reasons. The problem has not gone unnoticed. In a recent attempt to dispel these attacks, Evan Fales remarks for instance: "One source of dissent about the given has been a failure of those who employ the notion to be sufficiently clear and precise about what it is that they mean to assert when they speak of something as given" (Fales 1996, 6). And in his 1956 lectures entitled "The Myth of the Given: Three Lectures on Empiricism and the Philosophy of Mind" and subsequently published as the famous essay "Empiricism and the Philosophy of Mind" (Sellars 1956), which is classically credited for giving currency to the expression (Chisholm 1964, 128), Wilfrid Sellars describes his project as an effort to clarify and assess a whole series of criticisms of "the framework of givenness." He is thus led to distinguish different versions of the myth of the given before setting up his own one, although this version was in turn criticized for lack of precision and clarity.[6] A review of the different forms attributed to the myth of the given is thus an indispensable preliminary to any assessment of the vulnerability of Husserlian phenomenology to it. Unfortunately, this requirement cannot be fulfilled here, and there is no available study sufficiently comprehensive that might be used as a substitute.[7] I will therefore simply offer an outline of the main stages of the development of the myth of the given, as

6 See, for instance, Alston (1998).

7 Sosa (1997) provides however an interesting first step in that direction.

well as a rough set of criteria for the classification of the various versions of the myth resulting from it, before focusing on the specific one offered by Thomas Kuhn, which I see as being of special interest for the problem under consideration.

The starting point of this development is in fact contemporary with Husserl's phenomenology and can be located in the debate about the nature and function of protocol statements that divided the logical positivists. The famous criticism, expressed by Neurath (1932) of Carnap's idea that empirical science rests ultimately on indubitable protocol statements referring to immediate observation (Carnap 1932) is explicitly presented as a challenge to the claim that there are statements whose content includes only empirically given elements, and whose justification is indisputable in virtue of being immediate. The more protracted debate generated by the earlier publication in 1929 of C. I. Lewis's *Mind and the World-Order* can be seen as a second important stage, more directly centred on the existence of an empirical given than on the possibility of providing it with a pure propositional expression, as well as of making it the cornerstone of a foundationalist approach to empirical science. By defending what Frederick Firth (1949) later called a sensory core theory, Lewis was in fact far from being a radical adversary of the idea that something is given in empirical experience. He only rejected as misconceived the thesis that the given is immediate or "pre-analytic," and also that it is the correlate of an independent component of perceptive activity. Indeed, by arguing that every act of perception includes indissociable sense-data only disclosed by analysis, he himself came to be seen as a representative of a mythical conception of phenomenological properties by critics like John Wild (1940), or some partisans of what Roderick Chisholm also called the percept theory of perception. The publication of Wittgenstein's *Investigations* opened a third important phase, especially in relation to the argument against the possibility of a private language contained in the book. This argument has generated a puzzling number of interpretations and further elaborations, with the effect of dramatically increasing the number of versions of the myth of the given. According to what might be called its basic interpretation, well summarized by R. O. Jones (1971), its essential goal is to question the idea that consciousness gives mental objects of a private sort, in the sense of objects uniquely given to the subject of consciousness, such as pain, on the pretext that it would be in fact impossible to talk about these objects because of the nature of language. The problem is therefore concentrated here again on the very existence of the given, as well as on the possibility of its expression, and not on its enrolment in favour of an empiricist version of foundationalism. However, this aspect came again to the forefront in an important later phase of the development of the myth, namely the controversy between Wilfrid Sellars (1956) and Roderick Chisholm (1964). In Chisholm's definition, probably one of the more precise to be found in the literature, the existence of an empirically given is not even in dispute. Indeed, Chisholm simply defines the myth of the given, which he rejects, as the denial of two theses: (1) propositional knowledge rests on self-justifying statements, and (2) there are self-justifying statements about the given. Two additional episodes should at least be included in this more than sketchy historical account. One is the critique of the distinction between observation and theory put forward in the early sixties by post-positivist philosophers of science, notably by Thomas Kuhn (1962). This is more an epistemological debate than a cognitive one, although it is far from being without consequences on the analysis of cognitive structure, and of experience in particular. The second one, born in the early eighties and still alive, is on the contrary primarily cognitive in nature and deals with the reality of non-conceptual content. Although a non-conceptual content is not necessarily something given, an important criterion proposed for demarcating the given from the non-given, especially at the empirical level, has always been the absence of conceptual elements.

In spite of its complexity, it thus clearly appears that the denunciation of the myth of the given has constantly revolved around a limited number of issues raised in different contexts, as well as around a limited number of core ideas developed in different manners and combinations and supported by renewed arguments. Accordingly, its basic structure can be seen as composed of three negative claims to the effect that:

1. nothing is given to the human mind, in the sense that the human mind is never put in relation, in particular empirically, with something that can be considered as free from any contribution of its faculties;

2. what is possibly given to a human mind cannot be apprehended and / or expressed as such;

3. what is possibly given to the human mind cannot play the role of a foundation of human knowledge in the sense of providing an indisputable and absolutely basic knowledge.

2.2. The Kuhnian Version: Passiveness and Intuition

In the history of the critique of the given, Kuhn's analyses can be thought of special significance for several reasons. On the one hand, he gave exceptional historical importance to what he saw as mythical beliefs about the given since he considered them as "a traditional epistemological paradigm" initiated by Descartes and with a still-powerful grasp on late-twentieth-century reflection on knowledge, instead of as a specialized doctrine held by a minority of philosophers.[8] In addition, his critique put the emphasis on the same point as the one raised by Dennett against Husserl, offering thereby an interesting opportunity for clarifying the nature of this objection and assessing its real impact on Husserlian phenomenology. Finally, this critique also brings into sharp focus a conceptual distinction between a mythical and a non-mythical notion of the given which is of great theoretical importance, especially with regards to the question under discussion.

I will therefore offer a detailed interpretation of it and then proceed to examine to what extent Husserlian phenomenology can be legitimately accused of falling into the errors that it denounces.

Kuhn's critique is developed in the context of an analysis of scientific observation and, more specifically, of the defence of the idea that scientific revolutions, understood as paradigm shifts, include sudden observation shifts. These "observation switches," as Kuhn also calls them, do not designate the obvious fact that the evolution of scientific ideas is marked by the emergence of new observations or discoveries.

---

8 It should nevertheless be underlined that the Wittgensteinians gave a similar historical importance to their own version of the myth.

For such discoveries are part of what Kuhn sees as "normal science" or intra-paradigmatic research and consist in laying bare additional observational features of objects that at best lead to correct their perceptual identity. By contrast, observation shifts occur during inter-paradigmatic episodes of research and are characterized by the fact that objects are suddenly endowed with a new perceptual identity. Indeed, Kuhn's claim is that, as a scientific community gives up one paradigm in favour of another, "scientists see new and different things when looking with familiar instruments *in places they have looked before*," (Kuhn 1962, 111) so that the world is perceptually – and above all visually – transformed. The paradigmatic case of these observational metamorphoses is provided by the alleged difference in what Galileo and Aristotle respectively saw when observing movements such as a falling stone or a body swinging back and forth at the end of a string. In the second case, for instance, Galileo saw a pendulum where, according to Kuhn, Aristotle could see no more than a "constrained fall" (Kuhn 1962, 123). Accordingly, "the immediate content of Galileo's experience with falling stones was not what Aristotle's had been" (ibid.). From Kuhn's viewpoint, making sense of these profound transformations in observational content requires a no-less-profound change in the notion that observation is a process that gives something to the cognitive subject.

As made clear in the last quotation, observation is indeed for him a form of "immediate experience"; and he also unambiguously assimilates immediate experience with a process through which something is "given," and furthermore, "immediately given," to the observer (Kuhn 1962, 125). Therefore, Kuhn in no way questions the existence of such a thing as the given, and it is in his opinion perfectly legitimate to investigate what is given in observation to a scientist such as Galileo or Aristotle, and more generally in perception to a cognitive system. Phenomenology broadly construed as a theory of phenomenological properties (as they were defined above), and even phenomenology more narrowly understood as a first-person descriptive investigation of those same properties, are far from being ruled out in his eyes. Indeed, he explicitly laments that no such thing is available in the case of historical observation shifts, implicitly admitting at the same time that it is in the case of individual perception (115).

What is therefore mythical about the given, in Kuhn's opinion, is not its very existence, but a certain view about its nature that fails to provide adequate accounts of epistemological phenomena disclosed by the history of science such as observation switches. And the mythical element in such a view is the idea that what we perceive or experience is the passive end product of the stimulation of a specific faculty, namely the physiological apparatus of the senses. The fundamental belief under attack is therefore the assimilation of givenness with passiveness, although not with absolute passiveness, since it was clearly always accepted that the sense apparatus brings in its own contribution and that the given is to that extent perceptually constructed. The important point, however, is that no other faculty, especially one of a conceptual and inferential nature, is thought to contribute to the perceptual process, and consequently that the given is not conceptually and inferentially constructed.

Such a theory of the nature of immediate experience is clearly challenged by observation switches occurring during paradigm shifts since the physical stimulus and the perceiving system (putting aside individual factors as too minor to be relevant) remain identical, while the given is profoundly transformed. And although emphasizing that the two phenomena are different in spite of the obvious links uniting them, Kuhn also sees further evidence for this challenge to the traditional view in the similar situations of visual ambiguity affecting non-observational perception, such as the familiar duck–rabbit Gestalt switch. In addition, the correlation of observational switches with paradigm shifts clearly speaks in favour of a determination of the given by elements specific to the nature of paradigms, although Kuhn professes to have no specific theory to offer as to their nature and *modus operandi*. Whatever they are, the given is in this perspective no longer something passively received through the operations of the sense apparatus, no longer intuited, to resort to a terminology familiar to the epistemological tradition. It is crucial to underline that Kuhn, however, in no way concludes to the necessity for rejecting the very idea that something is given in perception and in observation. But the notion of givenness is inescapably altered.

In what sense, then, does the given remain a given? Kuhn is unfortunately no more fully explicit about this than he is about the fact that his rejection concerns the equation of givenness with passiveness. However, here again he gives unambiguous indications that what in his eyes makes the given what it is is its immediacy. In formulating an objection to his own criticism, he reveals, for instance, that he takes the expression of immediate experience to refer to the "perceptual features that a paradigm so highlights that they surrender their regularities *almost upon inspection*" (Kuhn 1962, 125). In other words, from the assimilation of experience with something immediately given, Kuhn only retains the notion of immediacy, while discarding that of givenness in its traditional (and etymological) sense of passive reception, which again is the traditional notion of intuition. The given is not given in any proper sense of the word; it is only immediate. It is what the subject immediately apprehends by means of the cognitive operations it performs, and there is nothing mythical about it. Kuhn thus brings to light a *distinguo* between two ideas which were undisputedly often conflated: namely, those of intuition and immediacy. And on this point he is in fact rediscovering, in a somewhat more confused way, a thesis clearly, if succinctly, put forward by Carnap in his *Aufbau*.[9] Indeed, although Kuhn probably sees the Aufbau as the epitome of the traditional epistemological paradigm he is criticizing, upon closer reading, the 1928 book reveals that the given is in Carnap's eyes whatever is "epistemically primary," and consequently, that givenness means for him nothing else than epistemic priority, precisely in the sense of epistemic immediacy.[10]

Kuhn's analysis also contains some developments about a possible move to avoid rejecting the mythical element of intuitiveness from the given, which throw a useful light on what he sees as the real nature of this intuitiveness. One could very well conclude from the facts of observation and Gestalt switches that what is wrong about the traditional paradigm is not its concept of givenness, but, on the contrary, its categorization of the immediately apprehended as something given. One way of doing this examined by Kuhn is to claim that what we

9 Carnap (1967). On can also find even clearer anticipations of it in Lewis (1929) or Wild (1940).

10 I have developed this view in more detail in Roy (2003b).

apprehend immediately, such as the pendulum perceived by Galileo in the case of vision, is the result of a theoretical interpretation of what is really given to us in experience. In this perspective, Galileo does not see a pendulum, in the sense of intuiting one, although this is what he immediately apprehends or perceives when he observes certain natural movements; his immediate apprehension is simply not purely intuitive, but includes a conceptual categorization of a theoretical kind. Consequently, strictly speaking, Aristotle and Galileo see the same thing when looking at those movements, but they apprehend or perceive them differently by virtue of a different theoretical interpretation. Both what they see and what they apprehend is immediate, but only what they see is given, while what they apprehend is theoretically interpreted. There is a layer of stable and theoretically neutral observational data across paradigms, as well as a layer of stable and theoretically neutral data across perceptual switches, and these two layers are intimately connected, if not identical. They constitute the real given, and they are given in the fundamental sense of being intuited, or again, passively received. Moreover, a "neutral language of observation" can be devised in order to faithfully and rigorously express descriptions of this purely intuited. Finally, it should be added to Kuhn's analysis that the truly given cannot be immediate in exactly the same sense as what is apprehended (Galileo's pendulum, again, in the case of visual perceptive apprehension). For what Galileo immediately perceives when opening his eyes is the pendulum, not the visual data that he interprets as such, and which arguably can only be laid bare through a process of analysis. However, it is no less arguable that the notion of passive reception is indissociable from some form of immediacy: even if the giving process is a long and indirect one, there is an obvious sense in which at the end of it, when the given is actually given, it is immediately apprehended.

Against such a strategy of regress, trying to locate pure passiveness at a deeper level of experience, Kuhn accumulates various counter-objections. One of them is of particular interest for later confronting his version of the myth of the given with Husserl's views. It shows, firstly, that Kuhn does not dispute the idea that scientific knowledge involves a process of theoretical interpretation of immediate experience. He even sees it as a major element of normal scientific investigation. But he denies that the observation switches disclosed by revolutionary observation switches can be assimilated with it: "The process by which either the individual or the community makes the transition from constrained fall to the pendulum, or from dephlogisticated air to oxygen, is not one that resembles interpretation.... Rather than being an interpreter, the scientist who embraces a new paradigm is like the man wearing inverting lenses. Confronting the same constellation of objects as before, and knowing what he does, he nevertheless finds them transformed through and through in many of their details" (Roy 2003b, 122). Observational switches occur at the level of observation itself. However, Kuhn's answer also makes it clear that observation cannot be conceived as a sort of low-level theoretical interpretation, although it is a non-intuitive process, and in that sense an interpretive one. But this interpretive process is in his eyes of a different kind than theoretical interpretation: it does not presuppose a layer of intuited elements. There is "no experience that is in principle more elementary" (Roy 2003b, 128) than what he designates as immediate experience and corresponds for instance to Galileo's perception of a pendulum versus Aristotle's perception of constrained fall. In fact, he goes as far as saying that no ordinary sense of the notion of interpretation fits the non-intuitive process at work in observation and perception, and revealed by observational and gestalt switches.

In sum, instead of denying that what has been traditionally categorized as immediately given is so, while retaining on the other hand the traditional notion of givenness as intuitiveness, Kuhn prefers to maintain the traditional categorization but modify the content of the category of givenness. As a result, the immediately given is for him something immediately apprehended but not intuited, and in that sense something involving interpretation, although in a manner that in no way resembles the theoretical interpretation it is itself subject to in normal science, and that is left unspecified.

3. Testing Husserl's Phenomenology

If we assume, on the basis of Kuhn's analyses, that what is mythical about the given is nothing less than the traditional notion of intuition, it is hard to see how Husserlian phenomenology could not be a myth-eaten doctrine, since intuition plays a crucial role in it. Here again, a preliminary survey of what is to count precisely as Husserlian phenomenology is indispensable but impossible to provide within the limits of this essay, although there are in this case more studies available that can help circumvent this limitation.[11] It is nevertheless preferable to concentrate on the seminal version of Husserlian phenomenology provided by Husserl himself. The conclusions to be obtained will therefore only show that his particular form of phenomenology does or does not subscribe to the sole Kuhnian definition of the myth of the given. Their impact will nevertheless be a significant one, given the central character both of this form and of this definition. In addition, this confrontation of Husserl with Kuhn must be carried out at two distinct levels. It is necessary to determine whether Husserl makes room for the notion of intuition seen as problematic by Kuhn in his analysis of cognitive faculties on the one side and in his conception itself of the nature of phenomenological investigation on the other side.

3.1. The Principle of Intuitionism

As already suggested, it is apparently difficult to provide negative answers to this double interrogation. Indeed, Husserl made clear as early as 1913 that the "principle of all principles" of his theory is an intuitionist one, according to which the human mind is endowed with a fundamental capacity for intuition that takes both empirical and non-empirical forms, all of which serve as ultimate sources of information as well as of justification in the acquisition of knowledge. As a result, science takes essentially the form of an intuition-based axiomatics, where every proposition is grounded in intuition either directly or indirectly, by means of an inference from other propositions directly grounded in it. The specific discipline of phenomenology constitutes however an exception: it is entirely based on intuition in a direct manner and hence represents a purely descriptive sort of scientific investigation.

11 See, for instance, the classic Spiegelberg (1994).

Husserl's notion of intuition is in addition quite explicitly linked to a notion of givenness. In his eyes, the hallmark of intuition lies in the fact that the subject has a direct cognitive relation with an object, in opposition to an indirect one, as in the paradigmatic case of the symbolic apprehension of an object, where the subject is related to its object through an intermediary one. Accordingly, an intuited object is present itself, "in person," and in that sense given to the subject. In "originary intuition," the purest form of intuition, it is furthermore present "in flesh" (*leibhaftig*), a notion essentially intended to capture its additional temporal presence. In imagination, for instance, the object is present in person without being present in flesh, because it is not apprehended as being contemporary with the act of apprehension, as it is in the case of perception. In the first volume of his *Ideen* Husserl writes: "Empirical intuition, experience especially, is the consciousness of an individual object; inasmuch as it is intuitive, it turns the object into a *datum*; inasmuch as it is perceptive, it turns it into an originary *datum*; through it we are conscious of grasping the object in an 'originary manner,' in its corporeal ipseity" (Husserl 1913, §3; my translation). And nothing essential seems to distinguish this notion of givenness from the one ostracized by Kuhn: Husserl emphatically describes intuition as a passive phenomenon whereby the subject is affected by an entity radically different from itself, and in immediate fashion.

Accordingly, the prospects are apparently very dim that any claim for the introduction of a phenomenological level of investigation into cognitive science could borrow from Husserl's phenomenology without offering a solid rebuttal of Kuhn's attack on intuition. And Husserl is threatened by this criticism with regard both to the content of its phenomenological descriptions and to his conception of phenomenological investigation. The first threat is mainly located in the notion of transcendent intuition, which Husserl sees as a basic component of cognitive activity. Without it, his whole doctrine crumbles down, and little significance is left to claiming affiliation with it. The second is located in the notions of immanent as well as eidetic intuition, considering that phenomenology is defined as an investigation based on a specific kind of intuition that is both immanent and eidetic.

It is therefore necessary to give a closer look at the reality of this twofold threat and, if it is confirmed, to determine whether it is as inescapable as it seems to be or whether there is any possibility other than that of disputing Kuhn's conclusions for salvaging the idea of a Husserlian phenomenological claim in cognitive science.

3.2. *Transcendent Intuition and Cognitive Theory*

Transcendent intuition is in Husserl's terminology the intuition through which spatio-temporal objects are given; it has two forms, external and internal, which are respectively at the core of outer and inner perceptions. There are in fact two main reasons for suggesting that, appearances notwithstanding, these two forms are free from any commitment to the myth of the given as defined by Kuhn.

In the first place, a central and well-known affirmation of Husserl's theory of cognition is that perception is in fact never purely intuitive. It is a process in which intuitive acts are intrinsically mixed with non-intuitive ones, more specifically designated as "signitive" (*signitiven*) acts in the fifth of his *Logical Investigations* (Husserl 1900–01). The content of a perceptive process is accordingly analyzed as a set of intuited determinations surrounded by a set of non-intuited ones, in such a way that the perceived object as a whole is never fully intuited at a single moment in time. As non-intuited features fall in the scope of intuition, intuited ones fall out of it. If Husserl's analysis grants a crucial role to the idea of intuition, it is therefore also a limited one: transcendent intuition is no more than an abstract component of spatio-temporal perception.

If this salient aspect of his doctrine downplays its exposure to the myth of the given, it remains insufficient to eliminate all danger. For, like C. I. Lewis in his sensory core theory, Husserl still maintains that experience comprises intuitive elements, although non-independent ones. Nevertheless, there is an additional and more compelling reason for seeing his conception of transcendent intuition as being actually beyond the reach of Kuhn's critique. Indeed, it is well known that Husserl is also strongly committed to a form of transcendentalism. And in virtue of such transcendentalism, what is intuited by the subject is in fact also the result of constitutive processes taking place within that subject. Transcendental constitution does not mean that what is characterized as an intuition at a certain level of analysis is, at a deeper one, revealed to be a construction, but rather that intuition as such is a constitutive process. For Husserl, the subject constitutes the object as intuited, as something passively affecting it. Indeed, the most striking result of his transcendentalism certainly is that it drives him to combine the two elements that Kuhn assumes to be antagonistic. Far from seeing the idea that an object is passively received by the subject as opposed to the idea that it is constructed by this subject, Husserl contends that givenness in the sense of passive reception is a form of constitution.

It is crucial to understand that in doing so Husserl does not and cannot revert to the Kantian view that the process of intuitive constitution consists in the conceptual interpretation of something intuited at the level of sensibility. This view, which was shown to be strongly rejected by Kuhn as a form of the myth of the given, cannot make, of what is immediately perceived, something itself intuited. It dissociates the immediacy of perception from its intuitiveness. On the contrary, Husserl does maintain that what is intuited in perception is part of what is immediately apprehended, of the perceived itself, and at the same time that it results from a process of constitution.

Consequently, he sees this constitutive process as a non-conceptual one, making in fact his approach very close on this point to Kuhn's, in spite of retaining the association between givenness and intuitiveness. Furthermore, he also analyzes it in terms of interpretation and offers a real hypothesis as to what such an interpretation might consist in. Indeed, the core of the constitutive process is for him a preconceptual and antepredicative interpretation, and the opposition between the intuitive and the non-intuitive is the result of a difference within such interpretation. The most detailed exposition of this aspect of Husserl's theory is offered in his fifth *Logical Investigation*. Making it fully intelligible would require technical developments beyond the scope of this essay, and, having already proposed elsewhere an in-depth analysis of this quite delicate point (cf. Roy 1999), I will only briefly recall its most essential aspect. It is a general principle of Husserl's theory that a mental state is an interpretation (*Auffassung*) by a noetic component or act, of a hyletic component, resulting in the emergence of an intentional object. The outcome of this interpretation depends on the nature of its various elements, chief among which are the matter (*Materie*) or sense (*Sinn*) of the act, the quality of the act, the content of the hyletic material, and finally its own form. The opposition between intuitive and non-intuitive states is specifically related to this last aspect: intuiting something is nothing else than giving a certain form to the interpretation of a *hyle* by a *noesis*. As a result, from the point of view of the intuitive act itself (versus from that of a reflexive view on the intuitive act), only the intentional object can be said to be given; the hyletic material itself, in particular, cannot be said so in spite of being frequently designated as "hyletic data." The point is worth being emphasized as the source of givenness is often mislocated in the quality of the act, and the hyletic material is sometimes understood as the truly given, as in Hintikka's interpretation of Husserl.[12]

Not only is Husserl thus revealed upon closer inspection to be largely on Kuhn's side about the given, but he actually goes beyond him by maintaining the notion of intuition as passive reception and integrating it into a fully interpretive framework, whereas Kuhn sees it as antagonistic with the notion of interpretation and eliminates it totally.

3.3. *Immanent Intuition and Phenomenological Investigation*

These remarks apply equally to the notion of immanent intuition and, if correct, they free therefore Husserl's phenomenology from any commitment to Kuhn's version of the myth of the given at the epistemological level as well: in other words, the idea of a non-interpretive apprehension of an object is not built either in Husserl's conception of the nature of phenomenological description.

As a matter of fact, nothing in the distinction between immanent and transcendent intuition hinges on the principles just reviewed. These principles specify a structure that is common to all forms of intuition. Or so it seems, at least, for the truth is that Husserl was far less explicit about the inner workings of immanent intuition than he was about those

---

12 Cf. Hintikka (1995) and Roy (1998) for further criticism of this interpretation.

of transcendent intuition. It is however a natural conclusion to draw, since none of the specific features of immanent intuition appears to be incompatible with such principles. Indeed, immanent intuition is firstly characterized by the fact that it occurs independently of an association with non-intuitive acts and consequently that immanent perception is pure:[13] everything is given in it. Furthermore, everything is given in it as it is, excluding the possibility of error. The existence of immanent intuition is also uncovered by suspending the validity of transcendent intuition (the operation of transcendental reduction) and it itself discloses non-spatio-temporal entities of an internal kind, namely the pure mental. Finally, it is intrinsically connected with phenomenological investigation in the sense that it provides it with a specific domain as well as a specific source of information and justification for making claims about this domain. There is no apparent reason why each one of these characteristics could not be accommodated with the general idea that intuition is a specific form of interpretation of a *hyle* by a *noesis*. And the same is true for eidetic intuition, although in both cases a serious difficulty certainly is to determine what sort of thing could play the role of a hyletic material.

It is important to note, however, that claiming that immanent intuition does not carry any commitment to the notion of givenness put into question by Kuhn does not amount to saying that it can and must be retained in the context of contemporary cognitive science, but solely that it cannot be discarded under that pretext. As a matter of fact, there are numerous reasons in my opinion for rejecting it. One is the highly disputable idea that it is irrefutable, and another that every aspect of the mental falls within its reach and is therefore conscious.[14]

---

13 While a transcendent perception is composed of intuitive and non-intuitive components, an immanent perception only comprises intuitive components. There is therefore no difference between immanent intuition and immanent perception.

14 This is a somewhat outdated idea, in sharp conflict with the current notion of a "cognitive unconscious," although it has been recently defended in a very interesting manner in the context of cognitive science by Perruchet and Vinter (2002).

But the most crucial reason is that, being disclosed as what survives transcendental reduction, it is in full contradiction with naturalism, a point that received for instance insufficient attention in Varela's version of the phenomenological claim known as 'neurophenomenology.'[15]

3.4. *The Dispensability of Intuition*

In order to really provide relief to the partisans of such a claim, however, these comforting conclusions have to be complemented by the demonstration that Husserl's way out of Kuhn's criticism is acceptable. If it is not, one can legitimately infer that Husserl's phenomenology does little more than devise an ingenious but unsuccessful solution for escaping a central version of the myth of the given. The question is therefore quite important, but I confess that I have no definitive answer to propose for it, in particular because it calls for extensive investigations of different sorts. It is consequently interesting to consider at this point an alternative problem, namely whether the idea of a phenomenological investigation as previously defined could survive the complete rejection of that of intuition. If it can, the question of the acceptability of Husserl's solution will lose some of its urgency.

On the face of it, the link between the two ideas is an intrinsic one, and it is introduced through the further notion of description: to describe is to conceptualize and express something that is intuited as it is intuited. And this corresponds indeed to Husserl's understanding of descriptive knowledge. It is interesting to note that the connection of Husserl's phenomenological description with an intuitive base (that is, with immanent intuition) is often confused with a connection with an object of description of an intuitive kind. But to say that a phenomenological description is always based on an immanent intuition is not to say that what is given in that intuition, and therefore described, must itself be an intuition. In that sense, Husserl's phenomenology is in no way

15 Cf. Varela et al. (2000), as well as Varela et al. (1993) and Varela (1996) for a general presentation of neurophenomenology.

uniquely directed to intuitions, and it has equal interest in the different sorts of components of our cognitive life. This confusion is probably the source of the one about hyletic data mentioned above. Hyletic data are certainly intuited from the point of view of the immanent intuition supporting phenomenological description, but they are not from the point of view of the various mental states in the composition of which they enter, including intuitive ones such as external perception. The question to be considered here is thus whether a phenomenological description could be based on something else than an intuition, or if it necessarily calls for an intuitive basis.

In fact, it raises nothing less than the problem of the nature of the type of knowledge embodied in consciousness, since, according to the definition of it that was previously laid out, a phenomenological description is by nature a report of the content of what we know in virtue of being conscious. This problem is a difficult one and one also that has not received sufficient attention so far. There is indeed a wide if implicit consensus that consciousness, as far as it is understood as awareness, is one form or another of intuition: when I see an apple and am aware that I do, my knowledge that I do is immediate and similar to a perception, and in that sense intuitive, whatever the notion of intuition ultimately retained. My ambition is not here to assess the value of such a view, but only to explore the consequences on the notion of phenomenological description of maintaining the contrary one that consciousness is not intuitive, and to suggest that there is none. Imagine, indeed, that awareness is in fact either of a purely theoretical nature or at least permeated with theoretical knowledge so that, in the first possibility, what the perceiving subject obtains through consciousness about his act of perception and its objective correlate is a purely theoretical hypothesis. Nothing is thus given to him, in the sense of being intuited, although it is arguable that consciousness can very well retain its immediate character and that something is therefore given to him in the sense of being immediately apprehended by him. There is still a specific way things immediately look to him, something it is like to look at an apple, and consequently something to be reported as faithfully and accurately as possible. In such a perspective, what the phenomenologist reports is the immediate theorizing produced by consciousness, and there is no reason why it should not be viewed as a descriptive activity.

The general conclusion is therefore that the minimal notion of the given which is indispensable to the phenomenological claim presented above is the one accepted by Kuhn – givenness as immediacy – even though this claim seems to be also immune from the criticism that Kuhn directs to the other notion – givenness as intuitiveness. Such a claim also needs to make the related assumption that what is given in consciousness in the minimal sense of being immediately apprehended is, at least to a certain extent, faithfully reportable. And this is the difficulty at stake in Dennett's objection against Husserl's phenomenology; however, it has just been shown that, contrary to what Dennett believes, this difficulty is in no way specifically related to the idea of the given as something intuited, and *a fortiori* with the notion of givenness as an interpretation-free apprehension of an object.[16] For it would no less concern the reportability of consciousness conceived as a purely theoretical source of knowledge. It is indeed possible that, in our efforts to report what we are conscious of, we would systematically distort it by introducing theoretical assumptions not contained in the immediately apprehended ones. But if it is, the problem of the reportability of consciousness in Husserl is independent from the question of Husserl's allegiance to the myth of the given, and as a consequence, it should be carefully separated from it in order to receive adequate treatment.

---

16 Dennett no less clearly makes the error of misreading Husserl on immanent intuition by seeing it as unrelated to any form of interpretation, although this additional mistake plays no special role in his argument, which deals entirely with the reportability of consciousness

References

Alston, W. (1998). "Sellars and the 'Myth of the Given'," *Eastern Division Meeting of the American Philosophical Association.*

Carnap, R. (1932). "Die Physikalische Sprache als Universalsprache der Wissenchaft," *Erkenntnis* **3**: 432–465.

Carnap, R. (1967) *The Logical Structure of the World*. London: Routledge and Kegan Paul.

Chisholm, R. (1964/1982). "Theory of Knowledge in America." In R. Chisholm, ed., *The Foundations of Knowing*. Minneapolis: University of Minnesota Press.

Dennett, D. C. (1991). *Consciousness Explained*. Boston: Little Brown.

Fales, E. (1996). *A Defense of the Given*. Lanham, MD: Rowman and Littlefield.

Firth, R. (1949). "Sense Data and the Percept Theory," *Mind* **58**: 59.

Hall, H., and H. L. Dreyfus (1982). *Husserl, Intentionality, and Cognitive Science*. Cambridge, MA: MIT Press.

Hintikka, J. (1995). "The Phenomenological Dimension." In D. Smith and B. Smith, eds., *The Cambridge Companion to Husserl*. Cambridge: Cambridge University Press.

Husserl, E. (1900–1901). *Logische Untersuchungen*. Halle: Max Niemeyer. English trans. by J. N. Findlay, *Logical Investigations*, London: Routledge and Kegan Paul, 1973.

Husserl, E. (1913). *"Ideen zu einer reinen Phänomenologie und phänomenologischen Philosophie," Jahrbuch für Philosophie und phänomenologische Forschung*. Halle. English trans. by W. R. Boyce Gibson, *Ideas Pertaining to a Pure Phenomenology*, New York: Collier Books, 1972.

Jones, O. R. (1971). *The Private Language Argument*. London: Macmillan.

Kuhn, T. (1962). *The Structure of Scientific Revolutions*. Chicago: University of Chicago Press.

Lewis, C. I. (1929). *Mind and the World-Order: Outline of a Theory of Knowledge*. New York: Dover.

Neurath, O. (1932). "Protokollsätze," *Erkenntnis* **3**: 204–14.

Perruchet, P., and A. Vinter (2002). "The Self-Organizing Consciousness," *Behavioral and Brain Sciences* **25**: 297–388.

Petitot, J. P., F. J. Varela, B. Pachoud, and J.-M. Roy, eds. (1999). *Naturalizing Phenomenology: Issues in Contemporary Phenomenology and Cognitive Science*. Stanford: Stanford University Press.

Roy, J.-M. (1998). "L'intentionnalité au carrefour de la phénoménologie et de la tradition sémantique." In E. Rigal, ed., *Jaakko Hintikka*. Paris: Vrin.

Roy, J.-M. (1999). "Saving Intentional Phenomena: Intentionality, Representation and Symbol." In J. Petitot et al., eds., *Naturalizing Phenomenology: Issues in Contemporary Phenomenology and Cognitive Science*. Stanford: Stanford University Press.

Roy, J.-M. (2000). "La pertinence cognitive de la phénoménologie husserlienne," Paper delivered at the October 2000 session of the *Phenomenology and Cognition Workshop*. Paris.

Roy, J.-M. (2000/2). "Déficit d'explication et revendication phénoménologique." *Intellectica* **31**: 35–83.

Roy, J.-M. (2003a). "La référence phénoménologique." In J. Poust and E. Pacherie, eds., *La philosophie cognitive*. Paris: Ophrys.

Roy, J.-M. (2003b). "Carnap's Husserlian Reading of the *Aufbau*." In C. Klein and S. Awodey, eds., *Rudolf Carnap – From Jena to L.A.: The Roots of Analytical Philosophy*. Chicago: Open Court.

Sellars, W. (1956). "Empiricism and the Philosophy of Mind." In H.F.M. Scriven, ed., *Foundations of Science and the Concepts of Psychology and Psychoanalysis*. Minneapolis: University of Minnesota Press.

Sosa, E. (1997). "The Mythology of the Given," *The History of Philosophy Quarterly* **14:** 275–86.

Spiegelberg, H. (1994). *The Phenomenological Movement: A Historical Introduction*, 3d ed. The Hague: M. Nijhoff.

Varela, F. J. (1996). "Neurophenomenology: A Methodological Remedy to the Hard Problem," *Journal of Consciousness Studies* **3(4)**: 330–49.

Varela, F. J., E. Thompson, and E. Rosch (1993). *The Embodied Mind*. Cambridge, MA: MIT Press.

Varela, F. J., N. Depraz, and P. Vermesch (2000). "The Gesture of Awareness: An Account of its Structural Dynamics." In M. Velmans, ed., *Investigating Phenomenal Consciousness: New Methodologies and Maps*. Amsterdam: John Benjamins.

Villela-Petit, M. (1999). "Cognitive Psychology and the Transcendental Theory of Knowledge." In Jean Petitot et al., eds., *Naturalizing Phenomenology: Issues in Contemporary Phenomenology and Cognitive Science*. Stanford: Stanford University Press.

Wild, J. (1940). "The Concept of the Given in Contemporary Philosophy – Its Origin and Limitations," *Philosophy and Phenomenological Research* **1**: 70–82.

CANADIAN JOURNAL OF PHILOSOPHY
Supplementary Volume 29

Descriptive Phenomenology and the Problem of Consciousness

DENIS FISETTE

What is phenomenology's contribution to contemporary debates in the philosophy of mind? I am here concerned with this question, and in particular with phenomenology's contribution to what has come to be called the problem of (intentional) consciousness. The problem of consciousness has constituted the focal point of classical phenomenology as well as the main problem, and indeed perhaps the stumbling block, of the philosophy of mind in the last two decades (Fisette and Poirier 2000). Many philosophers of mind, for instance, Thomas Nagel (1974), Ned Block (1995), Owen Flanagan (1977), Colin McGinn (1991) and David Chalmers (1996), have acknowledged the properly phenomenological character of this problem; Nagel is even willing to entrust the study of phenomenal consciousness to what he calls an "objective phenomenology." Yet, the phenomenology to which these philosophers resort has little to do with the conceptual framework that was developed within the phenomenological *tradition*. They put forward an entity they term "phenomenal consciousness," but only in the hope that it may be explained by means of the theories that currently prevail in the philosophy of mind or in cognitive sciences. In this respect, their phenomenology should rather be understood on the basis of the classical notion of "disposition." The situation is different in the case of philosophers such as Hubert Dreyfus (1972), John Haugeland (1998), Barry Smith (1995), Jean Petitot (1995b), and Francisco Varela (1995), who explicitly associate themselves with the phenomenological tradition that is rooted in the works of Brentano and Husserl. At any rate, they appeal to a concept of phenomenology which is more adequately articulated, perhaps precisely because it hooks up to this tradition.

Yet, although better articulated, the term remains prone to confusion for at least two reasons. First, the phenomenological tradition includes Brentano and Husserl as well as Heidegger and Merleau-Ponty, that is, philosophers whose conception of phenomenology is as different, say, as the conception of mind defended by Hilary Putnam (1988) in 1960, i.e., functionalism, and the conception he endorses today. It goes without saying that the significance and relevance of a possible contribution of phenomenology to the problem of consciousness depends precisely on the way we characterize it. Secondly, the philosophical presuppositions which are conveyed within the phenomenological tradition may differ from one protagonist to another. These presuppositions are, for instance, radically different in Husserl and in Heidegger. Furthermore, it seems hardly possible to reconcile them with those that underlie the naturalist positions advocated by the greater majority of philosophers of mind. In this respect, attempting to exploit the conceptual resources of phenomenology for the purpose of solving the problem of phenomenal consciousness is one thing, endeavouring to convert it to a determinate form of naturalism is another.

Hence, in order to avoid all confusion when using the term and in order to thus simplify the way in which phenomenology may be brought together with contemporary philosophy of mind, we will use the term to designate Husserl's conception of it. In our opinion, the field of Husserlian phenomenology which is relevant as regards current problems in the philosophy of mind is the one he indifferently called "descriptive psychology," "intentional psychology," and "phenomenological psychology." Understood in terms of descriptive psychology, and to the extent that it describes in purely conceptual terms what natural sciences explain by means of scientific laws and principles, phenomenology must be distinguished from psychology understood as a natural science and, in particular, from experimental psychology which arose in the stride of psychophysics. This is what explains this paper's interest in the idea of *description*, and in the question as to what is descriptive in phenomenology as well as to what is phenomenological in the use which is made of it within contemporary philosophy.

In order to answer these questions I will first recount the origins, in the nineteenth century, of the notion of descriptive psychology and identify some of the distinctive marks of Husserl's concept of description. I will then examine some aspects of the concept of description as it has been used in the other tradition following Wittgenstein's later work, focusing on the question as to whether the two concepts coincide. In the third part of the paper, I will briefly sketch the problem of consciousness with respect to the problem of the explanatory gap, and I will succinctly consider the conditions under which descriptive phenomenology may contribute to this problem.

The Origins of Descriptive Psychology in the Nineteenth Century

In the nineteenth century, the notion of description was commonly used by philosophers in order to distinguish philosophical psychology from psychology understood as a natural science. Contrary to what the post-Quinean use of the term implies, the attribute 'descriptive' in 'descriptive psychology' was not used in contradistinction to 'normative.' Rather, it pointed to an aspect of psychology that concerns conceptual analysis and the way in which phenomena are apprehended. In this respect, it is opposed to psychology understood as an explanatory science, that is, to this other aspect of psychology which concerns the explanation of phenomena by means of causal and natural laws. As Dilthey pointed out in his 1894 paper "Ideas concerning descriptive and analytical psychology," this distinction goes back to Christian Wolff and is taken up by Herbart as well as by Wilhelm Wundt.[1] Although Dilthey does not provide a clear definition of descriptive psychology, he refers the reader to the work of his Berliner colleague, Carl Stumpf, who was himself a student of Brentano.[2] But the point of reference as regards description, at least for Husserl and the other early phenomenologists,

---

1 See the introduction to his *Grundzüge der physiologischen Psychologie* (Leipzig: Engelmann, 1902), 1–10.

2 The reference to Stumpf's *Tonpsychologie* is in W. Dilthey, "Ideen zu einer beschreibenden und zergliederden Psychologie," in W. Dilthey *Descriptive Psychology and Historical Understanding* (The Hague: Nijhoff, 1977), 21–121.

is without any doubt Brentano.[3] Although the phrase 'descriptive psychology' does not feature in his *Psychology from an Empirical Standpoint* (1874), Brentano often uses it from the mid-1880s, and it even provides the title for some of his lectures (1890/1). He also referred to it as "descriptive phenomenology" or "psychognosis." As Husserl explains in his "Erinnerungen an Franz Brentano," it is to Brentano's descriptive psychology that he owes his first philosophical insights, and psychognosis can be seen to constitute the primary source of his phenomenology. Although it took other names within his philosophy, descriptive psychognosis remains its steadiest concern, perhaps even its guiding thread, from the very first writings up until the *Krisis*. A one-sided and even sometimes mistaken interpretation of Husserl's critique of psychologism is one reason among others that explains that the close connection between phenomenology and descriptive psychology has been neglected. We may mention, for instance, that the presence of descriptive psychology in Husserl's first book is so considerable that we could very nearly call it a "Brentanian Philosophy of Arithmetic."[4] Let us also recall along with Robin Rollinger in his *Husserl and the School of Brentano*, that the *Logical Investigations*, the groundwork of phenomenology, may be

---

3 Brentano uses the same expression in "Vom Ursprung sittlicher Erkenntnis" (3 and 14) to characterize his own philosophical contribution, and he also refers to descriptive psychology as the "analysis of mental phenomena into their elements" in his 1892 lecture "Über die Zukunft der Philosophie" (79) when he condemns those who neglect the observation of phenomena when investigating their causal origins. This question is raised in a more direct manner in his 1894 text "Meine letzte Wünsche für Österreich" (34) in which he establishes a connection between the term "psychognosis" and the terms "geognosis" and geology. Brentano there describes the task of [descriptive] psychology as consisting in the determination of the common features of mental phenomena and the investigation "of the ultimate elements of the mental life" (63). The task of genetic psychology is to provide a causal and nomological explanation of the same phenomena. It is thus closer to physiology and [natural] sciences in general. The ultimate constituents which are investigated by [descriptive] psychology are the mental phenomena which Brentano divides in three classes, namely presentations, judgment and sentiments, to which we have access through internal perception.

4 I borrow this expression from David Bell, "A Brentanian Philosophy of Arithmetic," *Brentano Studien* **2** (1989): 139–44.

understood as the result of Husserl's confrontation with Brentano and the latter's other students on the question of immanent contents and intentional objects. In fact, not only is phenomenology therein defined in terms of descriptive psychology, but the six investigations of this monumental work are meant as a tentative application of descriptive psychology to the basic concepts of pure logic. It is true that Husserl will reconsider this definition immediately after the publication of the *Logical Investigations*, but this reassessment does not affect the close relation between phenomenology and descriptive psychology. This is indeed confirmed by the three volumes of the *Ideas* in which Husserl defends the idea of a parallelism between intentional psychology and transcendental philosophy, and this idea is indeed taken up and systematically worked out in the mid-1920s as psychology again becomes the focus of his interest. During this period, phenomenology is defined equally in terms of intentional psychology and philosophy. In his Amsterdam lectures, for instance, Husserl claims that psychology differs from philosophy only by a tinge. Those who would be tempted to see in this an exceptional episode in the development of Husserl's thought should consider the manuscripts from the late 1920s[5] in which Husserl, in the context of the English translation of *Ideas I*, endeavours to refocus this work on intentional psychology. Psychology, insofar as it provides it with its raw material, is then understood not only as a propedeutic to philosophy but also as its way in. Finally, we may note the numerous courses, manuscripts, and papers devoted to psychological investigations during this period and remind ourselves that psychological investigations monopolize more than a third of Husserl's last work, the *Krisis*.

For the phenomenologist, the objects of description are the genuine modes of givenness of phenomena. Phenomenology thus stands in contrast to a certain form of phenomenalism. These genuine data correspond, in Husserl, to what Brentano calls the data of internal perception and, to the extent that they are immune to mistake and doubt, are immediate, etc., they have a status similar to the latter.

5 Those manuscripts were published by Karl Schuhmann in his edition of the *Ideas I* (*Hua III*).

They also have an epistemic function: they provide evidence and thus the foundation on the basis of which Brentano sought to justify the distinguished philosophical privilege which falls to psychology. And the comparison to Brentano's descriptive psychology does not stop there. Although 'analysis' takes a different sense in Husserl, the task which ensues to descriptive psychology consists in both cases in *analysing* "the totality of ultimate psychological elements." At any rate, we usually agree in saying that descriptive psychology represents a preliminary and unavoidable step towards scientific psychology. In a few words, Husserl agrees with Brentano on the idea that analysis of phenomena, their decomposition and composition, requires the intuition (*Einsicht*) of universal and apodictic laws. At any rate, one must admit that, although he also brings about important modifications, Husserl's intentional psychology follows Brentano's ideas on many significant points. The best-known of these modifications emerge from his critique of the doctrine of intentional inexistence in the 1890s and in the *Logical Investigations*; his critique of Brentano's dualism between internal and external perception in the sixth *Investigation*; his critique of the Brentanian notion of abstraction; and the recurring reproach as to the lack of an adequate method in Brentano. Husserl's amendments to Brentanian psychology are informed by his doctrine of essences, and it is in this perspective that eidetic psychology may ultimately be said to consist in a reform of Brentano's psychology. The chief differences between the two philosophers impinge on the heart of Husserl's philosophical project, a project that is already at work in his *Logical Investigations* and that concerns the link between psychology and ontology. These few remarks should suffice to show the endurance of this Brentanian line of investigation in Husserl's work. By emphasizing the relation between phenomenology and descriptive psychology, I wish to suggest that it is precisely this aspect of phenomenology which is relevant as regards its relation to cognitive sciences and the philosophy of mind in general. For, in my opinion, we are dealing, from a strictly philosophical standpoint, with two fairly different things. On the one hand, we have, in the philosophy of mind and cognitive sciences, a species of philosophical naturalism which we could qualify, on the basis of the philosophical status which falls to natural sciences and in particular to psychology, as 'psychologistic naturalism' (Fisette and Poirier 2000). On the other hand, this species of psychologism constitutes the main target of Husserl in the *Prolegomena*, and he will reiterate his critique on many occasions in the course of his work even to the detriment of the basic role that falls to descriptive psychology within his phenomenology.

Fourteen Defining Features of Husserl's Concept of Description

It might be tempting to resort to the concept of intentionality as a descriptive criterion for phenomenology and to construe the latter solely in terms of intentional analysis. But, if we follow the first edition of the *Logical Investigations*, 'descriptive content' and 'intentional content' are not equivalent and, indeed, Husserl's early phenomenology, whilst dubbing itself descriptive, is mainly concerned with the real constituents of experience, that is, to sensorial contents.[6] This is confirmed by Husserl's analyses of perception in the sixth *Investigation*. Husserl there claims that one of the characteristics of the (descriptive) essence of perception is that it "gives the object" (*LI* VI, 685), that it consists in the fact of having thus and thus impression, but that, as such, it is not intentional. It owes its intentional character to an act of *apprehension*. It follows from that that the modes of givenness which seem to be favoured by phenomenology and about which we have said above that they make out its descriptive character, instead of being restricted to intentional experience, as is often claimed, should rather be identified with the immanent content of experiences such as sensorial data and pictures.[7]

6 In a note from the second edition of the fifth *Logical Investigation*, Husserl notes: "The word 'phenomenological' like the word 'descriptive' was used in the First Edition only in connection with real [reelle] elements of experience, and in the present Edition it has so far been used predominately in this sense. This corresponds to one's natural starting with the psychological point of view"(*LIV*, 576).

7 In *Ideen*, §§85–86, Husserl asks again the question of non-intentional experiences and sensations, what he from now on will call *hylè* and which he connects with

Let us now take a look at some of the defining features of Husserl's notion of description, keeping in mind that the phenomenologist's interest in descriptive psychology lies chiefly in its descriptive character. We might want to start with the following analogy which will first allow us to demarcate descriptive and genetic psychology.

1) Descriptive psychology is to genetic psychology what natural history is to the natural sciences.

Descriptive psychology is the study of mental contents or experiences and its task, just like geognosis or natural history, is to analyse and classify (see *Ideen III*, 73 and Mc 124). By contrast, genetic psychology provides an explanation of "their origin and perishing, and the causal patterns and laws of their formation and transformation" (*LI* V, 545–46).

2) Phenomenological descriptions are not about the physical properties of empirical experiences but about the essence of these experiences.

In other terms, phenomenology is primarily concerned with the way in which we experience things, their modes of givenness or their appearing, not by what is given or appears, that is, by the properties of things. The analysis of the latter pertains to natural sciences. Hence, the empirical

---

Stumpf's notion of appearance. He adds that Stumpf's phenomenology is a doctrine of appearance or primary contents and that it corresponds to the hyletic and writes that "the idea of the hyletic eo ipso is transferred from phenomenology to the basis of an eidetic psychology which, according to our conception, would include Stumpf's 'phenomenology'" (*Ideas I*, 210). This passage confirms what he says of phenomenology in the *Logical Investigations* (see previous note). It also confirms that this phenomenology as well as that of Stumpf pertain to eidetic psychology. This is relevant to the extent that there is a connection between the *hylè* and the consideration that pertains to "phenomenal consciousness" in the philosophy of mind. In what follows, I will not however take into account what Husserl says about passive synthesis, although it has a direct impact on the questions with which we are dealing here.

concepts of content and of experience which, for instance, Wundt construes in terms of "event" disagree with the phenomenological concepts since the latter "cuts out all relation to empirically real existence" (*LI* V, 537). Take colour, for instance. We distinguish, on the one hand, the sensation of colour and the objective colour of the object, and, on the other hand, the experience from what is perceived, experience being here the sensorial moment of colour while the perceived is the object of the act of perception.

3) The phenomenological concept of phenomenon is not phenomenalist.

Husserl criticizes phenomenalist theories for confusing or neglecting the distinction between "appearance [*Erscheinung*] as intentional experience, and the apparent object (the subject of the objective predicates), and therefore identify[jng] the experienced complex of sensation with the complex of objective features" (*LI* V, 546). To put it plainly, phenomenalism ignores the important distinction, within the phenomenon, between, on the one hand, the object that appears and the appearing of the objects or experience and, on the other hand, the "real" constituent of experience (the sensorial content or picture) and its ideal constituent (the intentional content). Phenomenalism is thus liable to the same mistake as those who, like Brentano for instance, failed to differentiate the experienced from the perceived and who conceived of the latter as an event (thus implying that we perceive sense data).

Thence the fourth characteristic:

4) There is a distinction between intentional experience and immanent or non-intentional experience.

Husserl distinguishes immanent and intentional experience: he acknowledges that the "truly *immanent contents*, which belong to the real make-up (*reellen Bestande*) of the intentional experiences, are *not intentional*: they constitute the act, provide necessary *points d'appui* which render possible an intention, but are not themselves intended, not the objects presented in the act" (*LI* V, 559).

One does not see colour sensations but coloured objects, one does not hear auditory sensations but a melody, etc. Properly speaking, there are no immanent *objects*, there are only *intentional* objects. In the fifth *Logical Investigation*, Husserl often avails himself of an argument that rests on the possibility of error in the case of misleading perceptions. In these cases, according to Husserl, experience is the same whether the object exists or not: "It makes no essential difference to an object presented and given to consciousness whether it exists, or is fictitious, or is perhaps completely absurd," provided, as Husserl reminds in a note, that we abstract from the "various possible assertive traits involved in the believed being of what is represented" (*LI* V, 559).

5) In the early phenomenology, what constitutes the descriptive criterion for the distinction between mental and physical phenomena is not intentionality but rather experience understood in the wider sense.

As we have already noted, the only significant descriptive criterion for the two classes of phenomena (mental and physical) is the "descriptive character of the phenomena as they are experienced."

6) We do not access experience through internal perception as Brentano assumed; there is no difference between internal and external perception.

As his radical critique of Brentano's distinction between internal and external perception shows, Husserl's phenomenology is not introspective. There is, properly speaking, no difference between internal and external perception if by this we mean a difference between what is evident and what is dubitable. For even self-perception (as empirical self) is not, strictly speaking, evident (*LI* VI, 859). Because he confuses the experienced and the perceived, Brentano is led to claim that a phenomenon can be both a mental content (sensorial or pictural) and a mental object, and thus that consciousness can simultaneously and without modification be directed towards its object and towards itself.

7) There is an intuitive (non-referential) first-person mode of givenness which extends to the intuition of essences.

This is the doctrine of intuition of essences which Husserl develops in the sixth *Logical Investigation* and which I will not discuss here.

> 8) However, the privilege which falls to the first person should not be understood in the sense of a private and solipsistic experience; it is rather a standpoint or a situation.

This means that we must take into account the context of description. On one side, all description depends on the situation of the agent who is describing, that is, her body here and now, etc.; on the other, this situation is relative and makes sense only with respect to other situations, to a phenomenal field whose structure is comparable to a system of space-time coordinates.

> 9) The upshot of the first-person standpoint is a certain attitude which Husserl calls, in *Ideen II*, the personalist attitude.

Husserl also speaks of modes of apprehension: the naturalistic attitude which is characteristic of the theorist of natural sciences, and the personalist attitude in which one finds oneself whenever one is performing everyday actions. This attitude which Husserl also calls the "attitude of motivation" (*Ideen II*, 267) is characteristic of the mode of apprehension of the philosophy of mind. To these two modes of apprehension correspond two different objects: the correlate to the naturalistic attitude is the natural object (the appearing object) which can be explained by means of causal laws. By contrast, the correlate to the personalist attitude is the person understood as "subject of an environnement of persons and things." One and the same object, for instance one and the same behaviour, may be explained in many different ways; the appropriateness of the explanation depending directly on the way this behaviour is described or apprehended. When it is described or apprehended as natural, it lends itself to causal explanation. By contrast, when the same behaviour is described as intentional, the explanation will seek to elucidate its motivations.

> 10) The personalist attitude has precedence over the naturalistic attitude

Phenomenology grants ontological precedence to the personalist attitude, emphasizing that the naturalistic attitude is auxiliary at least in the following sense: not only does the latter presuppose the personalist attitude, but the naturalistic attitude comes about precisely when one abstracts from all intentions which are of the nature of sentiment and praxis.

11) The precedence of the personalist attitudes implies that natural explanation is merely derivative.

The idea, here, is simply that natural explanation is formed by leaving aside intentional descriptions and, above all, the manner in which the thing described is originally given.

12) The thesis of the irreducibility of phenomenological to natural description ultimately rests on a division of reality into regions.

Any attempt to reduce one type of description to the other is liable to a *metabasis allo genon* and to a category mistake. This Husserlian thesis is implied by the division made, within reality, between regions or regional ontologies, the two most important of which are the regions of nature and spirit (see *Natur und Geist, Husserliana XXXII*) (volumes of the *Husserliana* are henceforth cited as *Hua*, followed by the volume number).

13) The method of descriptive psychology is inductive and it is carried out in the natural attitude.

This method is characteristic of what one might want to call bottom-up phenomenology, which starts from the manner in which a phenomenon is originally given. It is elaborated in Husserl's lectures throughout the 1920s, and in particular in *Natur und Geist* (*Hua XXXII*) and *Phänomenologische Psychologie* (*Hua IX*).

14) Descriptive psychology constitutes a propedeutic to transcendental philosophy, but it is entirely distinct from it.

We have already tackled this question above and we will add here that, far from presupposing the transcendental reduction and other methodological devices which are meant to meet specific philosophical imperatives, it is rather the results of the descriptive activity of psychology which constitutes a precondition to transcendental philosophy.

The present list is not complete but it is sufficient for our present purpose. For what matters here is the fact that phenomenology understood as descriptive psychology provides us with a descriptive criterion restricting enough to decide what qualifies as phenomenological and what is truly descriptive in contemporary philosophy.

Some Uses of the Notion of Description since Frege

Ever since Frege, frequent use was made of the concept of description in analytical philosophy. The very idea of description is no stranger to what Frege calls "modes of givenness," a concept he uses as synonymous to sense or meaning. This concept furthermore owes its significance to the basic role played by meaning in our relation to objects or "referents." In particular, the emphasis on the modes of givenness of a behaviour aims at bringing to the fore the peculiarity of the relations and relations network which commands our trade with the external world. In this sense the notion of "mode of givenness" (*Gegebenheitsweise*) plays a role similar to that of description. By "mode of givenness" one must understand the way in which an act intends or aims at its object. It allows one to do without the agent's perspective in her relation to the object, that is, to the knowledge she has of the referent which is always only incomplete and which appreciably varies from one speaker to another. Two points may be raised in favour of meaning and modes of givenness in our reference to objects. First, the irreducibility of the speaker's knowledge of the meaning of an expression to the knowledge of this expression's reference or denotation. For, as Frege (1980) explains, if the knowledge of meaning could be reduced to the knowledge of reference, we would always be in a position to ascertain, for instance, that Hesperus and Phosphorus designate the same referent. Yet, as we just noted, since our knowledge of the object is always only incomplete, this is not the case. The second reason concerns what Wittgensteinians have called the "cognitive content" and which is conveyed by a sentence which we hold true. Frege asks how a sentence of the type "Phosphorus is Hesperus" could convey any information or teach us anything if its meaning could be reduced to its referent (to its truth-value). For if the sense of the expressions that turn up on each side of the identity symbol could be reduced to what these expressions have for reference, then a speaker who understands the meaning of this sentence, if he also holds it true would know by this very fact that these expressions stand for the same object and the sentence would impart him with no new information. In other words, there would be no difference between "Emile Ajar is Romain Gary" and the tautology "an author is an author." One must thus assume that the speaker who understands the meaning of these expressions may not know that the terms of this relation refer to the same object. Obviously, the idea of description throws into light one of the essential properties of what we may want to call act contexts, the context of belief, desire, intention, for instance; or referential opacity, which in many respects constitute the very trademark of phenomenological description. More on this later.

The concept of description used in analytical philosophy after Frege has often been connected with classical phenomenology. We may think of Russell's (1904) definite descriptions which he introduced in reaction to Meinong's theses on intentional objects. We may also evoke Wittgenstein's last writings on the philosophy of psychology whose main interlocutors were precisely those who defended descriptive psychology from Brentano to gestalt psychology. It is possible to show that these texts of Wittgenstein entail numerous more or less explicit references to concepts which are indissociable from the descriptive psychology of Brentano and his school, that of description in particular, which he uses extensively in connection with the phenomenon of gestalt-switch. Moreover, it is Wittgenstein's students who, in reaction to behaviourism and the neopositivist explanation of behaviour in nomologico-deductive terms, turned the notion of description into an operative concept of action theory and the philosophy of mind. In her book *Intention*, Elisabeth Anscombe used the expression "under a description" in order to throw into light a presupposition shared by all explanations of behaviour, namely the importance of the description which is made of it. Anscombe showed that there exists a correlation between the manner in which a behaviour is described and the manner in which it is explained. Similarly, other Wittgensteinians have argued that whenever behaviour is described in terms of intentional action, nomologico-deductive explanations are no longer appropriate because they violate what we elsewhere called the homogeneity constraint (see Chisholm 1964 and Anscombe 1979). Georg von Wright, another of Wittgenstein's students, puts forward a similar idea when he uses Aristotle's practical syllogism as a description and explanation schema of rational action.[8]

The descriptive feature inherent to the observation of behaviour was taken into account by philosophers such as Quine, Davidson, and Dennett, who advocate an "intepretationist" position in the philosophy of mind. Some aspects of interpretationism are related to descriptive psychology understood in the wider sense; in particular, Daniel Dennett's intentional stance and to the same extent, although for different reasons, Davidson's anomalous monism. Indeed, one could see Davidson's monism as an attempt to reconcile, precisely by means of the concept of description, Wittgensteinian anomalism with the deductive nomological model of the neo-positivists. In "Mental Events," Davidson grounds his anomalous monism on three apparently inconsistent principles:

P1 Causal interaction between the physical and the mental
P2 The nomonological character of causality
P3 The anomalism of the mental

The first principle deals with events in extension and is therefore "blind to the mental-physical dichotomy" (1980, 215). It simply states that some of the events described in psychological terms (mental events)

8 See Fisette and Poirier (2000, chap. 1). I take the opportunity to mention the dispute that opposes Searle (2000) and Dreyfus (2001) on the same question.

are the causes or effects of other events described in physical terms (physical events). The second principle states that every causal relation is necessarily nomological, that is, it is always subject to determinate laws. And the third principle states that events which are described in mental or psychological terms are not subject to laws. The two first principles seem to imply that action must be explained in nomological terms, what is however denied by the third principle – dear to the Neo-Wittgensteinans. But according to Davidson, the apparent inconsistency rests on the lack of a sufficient distinction between causal contexts, which are transparent, and nomological and explanatory contexts, which are opaque. Only the first of the three principles' context is transparent; the two others' are opaque since the way in which we refer to the terms of a nomological relation is relevant to the sentences that describe these relations (the nomological sentences). The third principle states that if one or both of the events which are connected by a causal relation are described in mental or intentional terms, the latter cannot be subject to laws. For Davidson (1980, 229), "an event is an action if and only if it can be described in a way that makes it intentional." Hence what turns behaviour into an action is solely the possibility of describing it in an appropriate vocabulary, namely as intentional. The same event may of course be described again in a vocabulary which is not intentional, the vocabulary of physics, for instance, and in this case it may be subject to laws. According to Davidson, the distinction to which Wittgenstein alludes does not concern the event itself, i.e., the *descriptum*, but only the description which is made of it in a situation of interpretation. Manifestly, in order to get rid of the apparent inconsistency between the three principles and thus resolve the conflict opposing Neopositivists and Neo-Wittgensteinians, it is sufficient according to Davidson to specify a description or a vocabulary within which two events which are connected by a causal relation cannot be subjected to laws: two causally related events can be subjected to laws *when they are described in a physical vocabulary*. When the distinctions between opaque and transparent contexts and between different modes of description are made explicit, the apparent inconsistency between Davidson's three principles disappears (see Fisette 1996).

Dennett is one of the few philosophers of mind to identify with (a certain) phenomenology, and this although he exhibits a certain propensity for eliminativism as regards the problem of consciousness (according to him, consciousness is one of those philosophical prejudices we have inherited from Descartes) and although he endorsed, from the very beginning, a determinate form of functionalism (beliefs, desires, intentions, etc., must be construed in terms of their functional roles). The relationship to phenomenology is to be sought in what he calls "stances," and more precisely, the intentional stance which constitutes the touchstone of his psychology whose task is to explain and predict behaviour. Roughly, the feature of his intentional psychology which allows him to characterize it as phenomenology is the emphasis on the idea of perspective and on a certain relativity of objects or referents to a conceptual framework. By insisting on this relativity, Dennett opposes a certain form of metaphysical realism, that which is advocated by John Searle, for instance, while moving towards a given species of interpretationism which is close to that of Quine or Davidson. Compared to his opponents or to the other proponents of interpretationism, Dennett's originality lies in the fact that whilst acknowledging a certain reality to the intentional system, what he also calls "real pattern," he admits that these "patterns" are only accessible to interpreters who adopt the appropriate perspective or attitude. An intentional system is thus only accessible and therefore ascribable to a given agent if such and such behaviour corroborates such and such interpretation. Among all those interpretations that are confirmed by a given behaviour, the best one will be the one that predicts best (1991: 45). In this respect, all explanation and all prediction depend on the description which is made. Dennett (1987) identifies three types of description and three corresponding attitudes: the "intentional stance," which corresponds to intentional description, the "physical stance," which concerns the physical state of the object in question and whose description is centred on our knowledge of the laws of nature and, finally, the "design stance," which rests on the notion of function and whose description concerns solely the functional aspect of the system described. The distinction between these three "stances" and the idea of correlative attitudes or descriptions (patterns) authorize Dennett's association to (a certain) phenomenology although, with his hetero-phenomenology, he breaks with most phenomenologists, whose introspectionism and the privilege they grant to the first-person standpoint he criticises.

The Problem of Consciousness and the Explanatory Gap

This being said, Dennett's skepticism towards consciousness and, more generally, towards phenomenal experience considerably distances him from the proper concerns of phenomenology as defined above. The version of functionalism he propounds in the philosophy of mind makes him liable to the extended criticism which Thomas Nagel and other philosophers who actually accept his arguments have however addressed to physicalism and functionalism.[9] In the opinion of Nagel, for instance, established theories, and physicalism in particular, do not have at their disposal the conceptual resources and the vocabulary which an adequate description of consciousness and of the mental in general requires precisely because of the subjective character of phenomenal consciousness: what it is like to be in such or such a state. Nagel and other philosophers mentioned above relate the question of consciousness and of phenomenal experience to properly phenomenological problems without, however, granting much importance to the specific character of phenomenology. The situation is different for philosophers who stem from the phenomenological tradition and who are concerned by the question as to whether and how phenomenology can contribute to this problem which, in many respects, is the stumbling block of all programs of naturalization in contemporary philosophy. Among the various theories suggesting that phenomenology should be integrated within a program of naturalization,[10] three deserve closer attention:

1. Petitot's program, which puts forward that this integration should be carried out via the mathematization of phenomenological descriptions;

---

9 For instance, the criticism Haugeland directs against to Dennett.

10 See Varela, F., J. Petitot, B. Pachoud, and J.M. Roy, eds. (1999). *Naturalizing Phenomenology: Issues in Contemporary Phenomenology and Cognitive Science*. Palo Alto: Stanford University Press.

2. Varela's program of neurophenomenology, which emphasizes the methodological aspect of phenomenology;
3. Flanagan's natural method, which proposes a division of labour and thus a collaboration between phenomenology, cognitive psychology, and neurosciences.

Because of the importance they bestow upon the problem of consciousness, I will chiefly address the programs put forward by Flanagan and Varela. But first, the question: what is the problem of consciousness? We know that, within contemporary philosophy, it first stood out in the mid-1970s and took the form of objections directed against physicalism and functionalism. The problem was the following: classical theories of the mind, and first and foremost functionalism, turned intentionality and mental representation into basic categories of the mind, thus hoping to list the whole set of defining features of the mind. This presupposition is precisely what those who resort to consciousness and think that it is resistant to a description and to an explanation in physicalist and functionalist terms dispute. In fact, virtually all objections raised against functionalism make use of the idea that consciousness is resistant to a functionalist explanation and to the other programs of naturalization of the mind we have at our disposal to this day. It goes without saying that the attitude of these philosophers as regards the very nature of this phenomenon differs considerably. We can nevertheless assert that their diagnosis of this problem largely confirms what Joseph Levine (1997) has termed the *explanatory gap*.

The problem Levine is seeking to pinpoint with this idea is the apparent asymmetry between, on the one hand, the sort of explanation which has for objects natural kinds such as water, which we may for instance describe by means of its chemical composition; and, on the other hand, the explanation, on the basis of the descriptive apparatus of functionalism or physicalism, of phenomenal consciousness. Insofar as the terms of the relation usually appear in identity statements such as "water = H_2O," we can here talk of explanations by identification. Such explanations are considered reductive because all properties of one of the terms, for instance water, may be reduced to the properties of the other, in this case those implied by its chemical composition.

In a reductive explanation, phenomena of a higher level are explained in terms of simpler entities or processes of a lower level. For instance, psychological capacities at the macroscopic level (e.g., memory, reasoning, etc.) are identified with properties of the brain at the microscopic level (e.g., properties of the neuronal network). The idea of an asymmetry also clearly emerges from the comparison between, say, the explanation of the natural phenomena of pain with, for instance, the functionalist explanation of pain in terms of "C-Fibre stimulation" or that put forward by Crick and Koch (1997) of a "neural oscillation of 35–75 hertz in the cortex." Let us compare explanations of the type: "water = H_2O" or "heat = level of activity of molecules" with the psychophysical statement: "pain = C-fibre stimulation." The *explananda* in sentences of the first type, water and heat, can be derived from microphysical facts and be explained on this basis. The same applies to biological phenomena such as reproduction, which may be explained by accounting for the cellular and genetic mechanisms which make it possible for an organism to produce other organisms, etc. Whenever this explanation satisfies certain conditions (whatever they may be) at the microscopic level, the phenomenon is indeed explained.

But this does not seem to apply to phenomenal properties. Indeed in the case of the identification of pain with, say, the stimulation of C-fibres, the relation is itself enigmatic if not entirely arbitrary. This is what many thought-experiments found in the literature reveal, Block's (1995) in particular. Those thought experiments are fit to show that it is possible for a creature to be in the state described by the *explanans* even when the state we seek to explain does not obtain.

According to Levine, the problem of consciousness understood as the problem of the explanatory gap concerns the nature of the phenomenon we seek to explain, the *explanandum,* as well as the conceptual framework on which the explanation is set up and the relation between the *explanandum* and the *explanans*. This diagnosis makes it thus possible for us to sharpen up the presentation of the problem:

1. Problems that concern the *explanandum*, i.e., all problems associated with the concept of consciousness. Some indeed think that the gap comes from the fact that science never actually provided

the concepts required in order to explain phenomenal consciousness.

2. Problems that concern the *explanans*, i.e., the conceptual framework on which the explanation must be constructed: physicalism, the different versions of functionalism, neurobiology or, more simply, as Rosenthal for instance assumes, non-conscious intentional states.
3. Problems that concern the relation *explanandum/explanans*. This problem depends of course on what we put in the *explanandum* and in the *explanans*.

The latter group of problems is especially important for the promoters of the explanatory gap who demand from a satisfactory explanation of consciousness that it deliver a complete conceptual analysis. As the following schema shows, functionalists construe this analysis in terms of function or causal role. The following schema represent the functionalist premises of a psychophysical identification:

1. Mental state M = the occupant of the functional role F (conceptual analysis)
2. The occupant of the role F = state of the brain B (empirical study)
3. $M = B$ (by transitivity)

Psychophysical identification presupposes the two operations that feature in the premises and which correspond, following Levine (1993), to the two stages of an explicative reduction. The first stage is the process which consists in shaping the concept of property which must be reduced, the *explanandum*, by identifying the causal role whose underlying mechanism we are seeking; the second stage is simply the empirical investigation of these underlying mechanisms. Although Levine and other philosophers deny the validity of functionalist analyses, they nevertheless acknowledge the necessity of developing a conceptual apparatus which is apt to provide an adequate analysis and an appropriate description of consciousness.

Phenomenology Faced with the Problem of Consciousness

Those who believe that phenomenology may be beneficial to the problem of consciousness think indeed that this descriptive device is already at hand and that it must only be extracted from the corpus of classical phenomenology. But whoever is even only slightly familiar with the latter knows for a fact that whilst intentional consciousness indeed constitutes a favoured topic for some phenomenologists, it is certainly not the central issue within phenomenology as a whole. On the contrary, it could be shown that many philosophers who associate with (a certain) phenomenology would rather adopt, as regards the problem of consciousness, a position close to that of philosophers of mind which see in it nothing more than a Cartesian prejudice. In other words, what applies to consciousness also applies to phenomenology: the latter concept designates fairly different things. Unless, of course, one associates it, as we did at the very beginning, with Husserl and restricts its possible contribution to the domain of intentional psychology – even if this means revising it if necessary, taking into account what was accomplished in this field after Husserl. For the sake of discussion, let us admit that we thus obtain a concept of phenomenology which is reliable and let us ask whether those who trust this concept to play a central role in their program are justified in doing so. If our diagnosis of the problem is correct and if it is true that the problem of the explanatory gap originates in a certain conception of the mind which some term "mentalist" and "representationalist," we must make sure that this phenomenology endorses a conception of the mind which is not liable to the same objections that we mentioned earlier. But, on this question, phenomenologists disagree. We may think of Heidegger's, but also of Merleau-Ponty's, objections against Husserl; we may also think of Dreyfus's thesis, which draws on these objections and according to which Husserl's doctrine of noema and of intentionality in general anticipates the central insights of classical computationalism and of functionalism and, for this very reason, engenders the same kind of problems as the latter such as that of phenomenal consciousness. As far as I am concerned, I acknowledge the validity of this critique of representationalism, and I do not conceal my sympathy towards what has recently been called situated cognition and to the conception of mind according to which thought, body, and environment form an organic and indissoluble whole. But I also believe that Husserlian phenomenology remains a plausible candidate to an adequate description of the *explanandum*, just like some theses of Merleau-Ponty, themselves drawing on Husserl, and ecological psychology.[11] This is what many theses associated above with Husserl's concept of description suggest. Such is, in any case, my working hypothesis and the main constraint which I impose on my assessment of Varela's and Flanagan's programs.

Flanagan's (1997) ecumenical approach, his "natural method," is attractive insofar as it takes into account the three aspects of the problem we have just identified, namely the *explanandum* (phenomenology), the *explanans* (neurosciences) and the relation between *explanandum* and *explanans* (psychology / cognitive sciences). Traditionally, each of these disciplines developed in adjusting its theories to new empirical data or in putting forward new theories which better concurred with these data. The natural method stipulates that the development of each of these disciplines must be constrained not only by its own empirical and conceptual data, but also by those of the two others. The chief rule of the natural method is to regard each of the three disciplines with an equal deference. The idea is simply to use phenomenology to inform the empirical inquest in psychology and neurology and to use physiology to inform psychology and phenomenology. Neuro- and cognitive-scientists should then take into consideration what phenomenologists say when describing the structure of conscious experience and should revise, if necessary, their theories and their conceptual framework so that they agree with phenomenal descriptions. But this rule also applies to phenomenologists who must take seriously what neuro- and cognitive-scientists say and must also revise their theories and

---

11 For a phenomonological description of perceptual experience enthused by ecological psychology, see E. Thompson (1995), chap. 6.

conceptual frameworks if need be. This is also the method put forward by Varela under the label of neurophenomenology. According to him, phenomenological descriptions of the structure of experience and their counterparts in cognitive and neuro-sciences are linked to each other by mutual constraints. These constraints are however very different from those imposed by Flanagan and that for reasons which I will now examine.

In the first place, there are important differences between Varela's and Flanagan's approaches as regards the three aspects of the problem of explaining consciousness. In fact, we are dealing with two conceptions of mind which are fairly dissimilar. Although Flanagan acknowledges the relevance of the problem of consciousness and attributes this problem to the lack of an adequate description of phenomenal consciousness, he never indeed puts into question the representationalist conception of mind. Moreover, the phenomenology to which he entrusts the study of phenomenal consciousness is at best metaphorical (it consists in the reports of a patient about his own qualitative or other types of states) and is only phenomenological by name. By contrast, the theory of enaction put forward by Varela, Thompson and Rosch (1991) commits them to a non-representationalist conception of mind, a conception close to that of Merleau-Ponty and of ecological psychology in which consciousness, understood in terms of phenomenal experience, plays a central role. The description of the structure of the latter pertains to neuro-phenomenology, whose method amply draws on classical phenomenology (Varela 1995). The question is now to explain in what sense the phenomenological description of the *explanandum* can be said to be truly constraining as regards the explanation of consciousness. For, a phenomenology that insists on the phenomenal dimension of experience is not necessarily compatible with the classical models that play the role of explanans in the philosophy of mind, in particular classical computationalism and connectionism. Varela and Petitot are right to resort to dynamical models, if only because the systems in which they are interested – systems that draw on the theory of dynamical systems – operate with a conception of mind which is compatible with such a phenomenology. Mind is therein understood not as a computer-brain isolated in the skull but rather as a fully-fledged organism which comprises a nervous system and a body situated in an environment.

The embodiment of mind, its dependence upon an environment and the importance of time as regards cognition (see van Gelder 1995) are all features that are peculiarly close to the dynamical approach to phenomenological descriptions of phenomenal experience. Those who, as Flanagan seems to do, endorse classical models, are forced to exclude any phenomenological constraints. Indeed, in the latter perspective, classical models are that which impose the norm.

Hence, there remains the third aspect of the problem of the explanation of consciousness which concerns the relation *explanans/ explanandum*. In *Naturalizing Phenomenology*, a book which brings together many philosophers who are sympathetic to phenomenology, three options are put forward which go in the sense of a naturalization of phenomenology and, to a certain extent, of a "phenomenologization" of the framework on which the explicative apparatus of the natural sciences rests. The first option is that of the "bridge locus," that is, the idea of a bridge between, for instance, the percept and the neuronal substrate which would take the following form: if an empirical event is similar to a phenomenal event, then the former explains the latter.

$$\Phi \text{ is similar to } \Psi \Rightarrow \Phi \text{ explains } \Psi$$

But this is not satisfactory, if only because, as we have said in connection to conceptual analysis, we demand more from an explanation. The second option is isomorphism, and it features all advantages and disadvantages of a theory of identity. Hence, there remains only what authors of the introduction of this book call "generative passages." Following this option which is endorsed by Varela (1997), we should turn to the concepts of morphodynamics as bridge-concepts between the (eidetic) concepts of a phenomenological description and those of geometry. This is a two-step move: first, from the phenomenological concepts to the morphological ones (self-organization); second from the morphological concepts to the geometrical ones (morphodynamics). According to the third option, it is dynamical theories which "make it possible to explain how small-scale (microscopic) units can get organized into large scale (macroscopic) emergent structures on the basis of coordinated (cooperative and conflicting) collective behaviour situated at an intermediary (mesoscopic) level)" (1999, 55). In other terms, the macroscopic level should emerge morphologically from the structures based on the peculiarity of microscopic processes. This option certainly raises many questions, both at the epistemological and the ontological level – we may think of the difficulties as regards the concept of emergence (see Kim 1998) – but, in my opinion, it nevertheless constitutes the most viable option as regards the possible contribution of phenomenology to cognitive sciences.

References

Anscombe, E. (1963). *Intention*. Oxford: Blackwell.

Anscombe, E. (1979). "Under a Description," *Nous* **13**: 219–33.

Bell, D. (1989) "A Brentanian Philosophy of Arithmetic," *Brentano Studien* **2**: 139–44.

Block, N. (1995). "On a confusion about the function of consciousness," *Behavioral and Brain Sciences* **18**: 227–47.

Block, N., O. Flanagan, and G. Güzeldere, eds. (1997). *The Nature of Consciousness: Philosophical Debates*. Cambridge, MA: MIT Press.

Brentano, F. C. (1874). *Psychologie vom empirischen Standpunkte*. Leipzig: Meiner, 1924.

Brentano, F. (1890/1). *Deskriptive Psychologie*. Hamburg, Meiner, 1982.

Chisholm, R. (1964). "The Descriptive Element in the Concept of Action," *Journal of Philosophy* **61**: 613–25.

Chalmers, D. J. (1996). *The Conscious Mind. In Search of a Fundamental Theory*. New York: Oxford University Press.

Crick, F. and Koch, C. (1997). "Towards a Neurobiological Theory of Consciousness," in N. Block, O. Flanagan, and G. Güzeldere (eds.), *The Nature of Consciousness: Philosophical Debates*, pp. 277–92. Cambridge, MA: MIT Press.

Davidson, D. (1970). "Mental Events." In D. Davidson, *Essays on Actions and Events*. Oxford: Oxford University Press, 1980.

Davidson, D. (1980). *Essays on Actions and Events*. Oxford: Oxford University Press.

Dennett, D. C. (1987). *The Intentional Stance*. Cambridge, MA: MIT Press.

Dennett, D.C. (1991) "Real Patterns," *Journal of Philosophy* **88**: 27–51.

Dilthey, W. (1977). *Descriptive Psychology and Historical Understanding*. The Hague: Nijhoff.

Dreyfus, Hubert L. (1972). *What Computers Can't Do: The Limits of Artificial Intelligence*. New York: Harper Colophon Books.

Dreyfus, H. (2001). □henomenological Description versus Rational Reconstruction, *Revue internationale de philosophie*, 181 96.

Dreyfus, H., and H. Hall, eds. (1982). *Husserl, Intentionality and Cognitive Sciences*. Cambridge, MA: MIT Press.

Fisette, D. (1996). "Davidson on Norms and the Explanation of Intentional Behavior." In M. Marion and R. S. Cohen, eds. *Logic and Philosophy of Science in Québec*, Boston Studies in the Philosophy of Science, Boston, Dordrecht: Kluwer. 139–57.

Fisette, D. (2002). "Erläuterungen: Logical Analysis vs. Phenomenological Descriptions." In R. Feist, ed. *Husserl and the Natural Sciences*, Ottawa, Ottawa University Press (forthcoming).

Fisette, D., ed. (1999). *Consciousness and Intentionality: Models and Modalities of Attribution*. Dordrecht: Kluwer.

Fisette, D., and P. Poirier (2000). *La philosophie de l'esprit: état des lieux*. Paris: Vrin.

Flanagan, O. (1997). "Prospects for a unified theory of consciousness or, what dreams are made of." In J. Cohen and J. Schooler, eds. *Scientific Approaches to Consciousness*. Hillsdale, NJ: Erlbaum.

Frege, G. (1980). *Translations from the Philosophical Writings of Gottlob Frege*. Edited by Peter Geach and Max Black. Oxford: Basil Blackwell.

Gibson, J. J. (1979). *The Ecological Approach to Visual Perception*. Boston: Houghton Mifflin.

Haugeland, J. (1998). *Having Thought: Essays in the Philosophy of Mind*. Cambridge, MA: Harvard University Press.

Husserl, Edmund, *Gesammelte Werke*. Auf Grund des Nachlasses veröffentlicht in Gemeinschaft mit dem Husserl-Archiv an der Universität Köln vom Husserl-Archiv (Leuven) unter Leitung von S. Ijsseling, Den Haag, 1950–, *Husserliana*. Dordrecht: Kluwer.

Kim, J. (1998). *Mind in a Physical World*. Cambridge, MA: MIT Press.

Levine, J. (1997). "On leaving out what it's like." In N. Block, O. Flanagan, and G. Güzeldere (eds.), *The Nature of Consciousness: Philosophical Debates*, pp. 543–56. Cambridge, MA: MIT Press.

McGinn, C. (1991). *The Problem of Consciousness: Essays Toward a Resolution*. Oxford: Blackwell.

Merleau-Ponty, M. (1942). *La structure du comportement*, 3d ed. Paris: Presses Universitaires de France, 1953.

Nagel, T. (1974). "What is it like to be a bat?" *Philosophical Review* **83**: 435–50.

Petitot, J., ed. (1995a). *Sciences Cognitives et Phénoménologie*. *Archives de Philosophie* **58** (Numéro spécial).

Petitot, J. (1995b). La réorientation naturaliste de la phénoménologie. *Archives de Philosophie* **58**: 631–58.

Putnam, H. (1988). *Representation and Reality*. Cambridge, MA: MIT Press.

Quine, W. V. (1960). *Word and Object*. Cambridge, MA: MIT Press.

Rollinger, R.D. (1999). *Husserl's Position in the School of Brentano*. Dordrecht: Kluwer.

Searle, J. (2000). "The Limits of Phenomenology," *Essays in Honor of Hubert Dreyfus*, vol. 2, Cambridge, MA: MIT Press, 71–92.

Smith, B. (1995). Normal Ontology, Common Sense and Cognitive Science, *International Journal of Human-Computer Studies* 43: 641 67.

Thompson, E. (1995). *Colour Vision*. London: Routledge.

Van Gelder, T. (1995). "What might cognition be if not computation," *Journal of Philosophy* **92**: 345–81.

Varela, F. J. (1995). "Neurophenomenology: A Methodological Remedy for the Hard Problem," *Journal of Consciousness Studies* **3(4)**: 330–49.

Varela, F. J. (1997). "The Naturalization of Phenomenology as the Transcendence of Nature. Searching for Generative Mutual Constraints," *Alter* **5**: 355–81.

Varela, F. J., E. Thompson, and E. Rosch (1991). *The Embodied Mind: Cognitive Science and Human Experience*. Cambridge, MA: MIT Press.

Petitot, J., Varela, F.J., Pachoud, B., and Roy, J.M. (eds.) (1999). *Naturalizing Phenomenology: Issues in Contemporary Phenomenology and Cognitive Science*. Palo Alto: Stanford University Press.

Wundt, W. (1902). *Grundzüge der physiologischen Psychologie,* Leipzig: Engelmann.

CANADIAN JOURNAL OF PHILOSOPHY
Supplementary Volume 29

Intentionality and Phenomenality: A Phenomenological Take on the Hard Problem

DAN ZAHAVI

In his book *The Conscious Mind* David Chalmers introduced a now-familiar distinction between the hard problem and the easy problems of consciousness. The easy problems are those concerned with the question of how the mind can process information, react to environmental stimuli, and exhibit such capacities as discrimination, categorization, and introspection (Chalmers 1996, 4; 1995, 200). All of these abilities are impressive, but they are, according to Chalmers, not metaphysically baffling, since they can all be tackled by means of the standard repertoire of cognitive science and explained in terms of computational or neural mechanisms. This task might still be difficult, but it is within reach. In contrast, the hard problem – also known as *the* problem of consciousness (Chalmers 1995, 201) – is the problem of explaining why mental states have phenomenal or experiential qualities. Why is it like something to 'taste coffee,' to 'touch an ice cube,' to 'look at a sunset,' etc.? Why does it feel the way it does? Why does it feel like anything at all?

Chalmers' distinction confronts us with a version of the so-called 'explanatory gap.' On the one hand, we have certain cognitive functions, which can apparently be explained reductively, and on the other hand, we have a number of experiential qualities, which seem to resist this reductive explanation. We can establish *that* a certain function is accompanied by a certain experience, but we have no idea *why* that happens, and regardless of how closely we scrutinize the neural mechanisms, we don't seem to be getting any closer at an answer.

In his book, Chalmers also distinguished two concepts of mind: a *phenomenal* concept and a *psychological* concept. The first captures the conscious aspect of mind: Mind is understood in terms of conscious experience. The second concept understands mind in functional terms as the causal or explanatory basis for behaviour. According to the phenomenal concept, a state is mental if it 'feels' a certain way; according to the psychological concept, a state is mental if it plays an appropriate causal role. The first concept characterizes mind by the way it *feels*, the second by what it *does* (Chalmers 1996, 11–12), and according to Chalmers it is the first concept that is troublesome and which resists standard attempts at explanation.[1]

In a later article from 1997 Chalmers seems to have modified, or at least clarified, his position slightly. He now concedes that such notions as attention, memory, intentionality, etc., contain both easy and hard aspects (Chalmers 1997, 10). A full and comprehensive understanding of, e.g., intentionality would consequently entail solving the hard problem, or to put it differently, an analysis of thoughts, beliefs, categorization, etc., that ignored the experiential side would merely be an analysis of what could be called pseudo-thoughts or pseudo-beliefs (Chalmers 1997, 20). This clarification fits well with an observation that Chalmers made already in *The Conscious Mind*, namely, that one could operate with a *deflationary* and an *inflationary* concept of belief, respectively. Whereas the first concept is a purely psychological (functional) concept that does not involve any reference to conscious experience, the second concept entails that conscious experience is required for true intentionality (Chalmers 1996, 20). In 1997, Chalmers admits that

---

1 Chalmers insists that the phenomenology and the psychology of mind are systematically related. The phenomenal structure is mirrored by the psychological structure and vice versa. Whenever there is conscious experience, there is also some corresponding information processing going on, and whenever there is information available in the cognitive system for control of behaviour, there will also be a corresponding conscious experience. Consciousness and cognition consequently cohere in a systematic and intimate way, and Chalmers speaks of an isomorphic relation, which he calls the *principle of structural coherence* (Chalmers 1995, 212–13; 1996, 218–25).

he is torn on the issue, and that over time he has become increasingly sympathetic to the second concept, and to the idea that consciousness is the primary source of meaning, so that intentional content may in fact be grounded in phenomenal content, but he thinks the matter needs further examination (Chalmers 1997, 21).

I welcome this clarification, but I also find it slightly surprising that Chalmers is prepared to concede this much. As far as I can see, the very distinction between the easy problems and the hard problem of consciousness becomes questionable the moment one opts for the inflationary concept. Given this concept, it seems natural to conclude that there are in fact no easy problems of consciousness. The truly easy problems are all problems about pseudo-thoughts, etc., that is, about non-conscious information processing, but a treatment of these issues should not be confused with an explanation of the kind of conscious intentionality that we encounter in human beings. In other words, we will not understand how human beings consciously intend, discriminate, categorize, react, report, and introspect, etc., until we understand the role of subjective experience in those processes (cf. Hodgson 1996).

Chalmers's discussion of the hard problem has identified and labelled an aspect of consciousness that cannot be ignored. However, his way of defining and distinguishing the hard problem from the easy problems seems in many ways indebted to the very reductionism that he is out to oppose. If one thinks that cognition and intentionality are basically a matter of information processing and causal co-variation that could in principle just as well go on in a mindless computer – or to use Chalmers's own favoured example, in an experienceless zombie – then one is left with the impression that all that is really distinctive about consciousness is its qualitative or phenomenal aspect. But this seems to suggest that with the exception of some evanescent qualia everything about consciousness including intentionality can be explained in reductive (computational or neural) terms; and in this case, epiphenomenalism threatens.

To put it differently, Chalmers's distinction between the hard and the easy problems of consciousness shares a common feature with many other recent analytical attempts to defend consciousness against the onslaught of reductionism: They all grant far too much to the other side.

Reductionism has typically proceeded with a classical divide-and-rule strategy. There are basically two sides to consciousness: Intentionality and phenomenality. We don't currently know how to reduce the latter aspect, so let us separate the two sides and concentrate on the first. If we then succeed in explaining intentionality reductively, the aspect of phenomenality cannot be all that significant. Many non-reductive materialists have uncritically adopted the very same strategy. They have marginalized subjectivity by identifying it with epiphenomenal qualia and have then claimed that it is this aspect which eludes reductionism.

But is this partition really acceptable, are we really dealing with two separate problems, or are experience and intentionality on the contrary intimately connected? Is it really possible to investigate intentionality properly without taking experience, the first-person perspective, semantics, etc., into account? And vice versa, is it possible to understand the nature of subjectivity and experience if we ignore intentionality? Or do we not then run the risk of reinstating a Cartesian subject-world dualism that ignores everything captured by the phrase "being-in-the-world"?

In the following, I wish to consider some arguments in favour of opposing the separation. I will try to supply some answers to the three following questions:

1. What forms of intentionality possess phenomenal features?
2. Do all experiences possess intentional features?
3. If the intentional and the phenomenal go hand in hand, is the connection then contingent or essential?

All three questions call for quite substantial analyses. All I can do in the following is to provide some preliminary reflections; reflections that will incidentally suggest that analytical philosophy in its dealing with these questions might profit from looking at some of the resources found in continental phenomenology. Why? Because many of the problems and questions that analytical philosophy of mind are currently facing are problems and questions that phenomenologists have been struggling with for more than a century. Drawing on their results would not only help avoid unnecessary repetitions, it might also bring the contemporary debate to a higher level of sophistication.

1. Is there a 'what it is like' to intentional consciousness?

It is relatively uncontroversial that there is a certain (phenomenal) quality of 'what it is like' or what it 'feels' like to have perceptual experiences, desires, feelings, and moods. There is something it is like to taste an omelette, to touch an ice cube, to crave chocolate, to have stage fright, to feel envious, nervous, depressed, or happy. However, is it really acceptable to limit the phenomenal dimension of experience to *sensory* or *emotional* states alone? Is there nothing it is like simply to think of (rather than perceive) a green apple? And what about abstract beliefs: is there nothing it is like to believe that the square root of 9 = 3? Many contemporary philosophers have denied that beliefs are inherently phenomenal (cf. Tye 1995, 138; Jacob 1998; O'Shaughnessy 2000, 39, 41). I think they are mistaken.

Back in the *Logical Investigations* (1900–01), Husserl argued that conscious thoughts have experiential qualities and that episodes of conscious thoughts are experiential episodes. In arguing for this claim, Husserl drew some distinctions that I think are of relevance in this context. According to Husserl, every intentional experience possesses two different but inseparable moments. Every intentional experience is an experience of a specific type, be it an experience of judging, hoping, desiring, regretting, remembering, affirming, doubting, wondering, fearing, etc. Husserl called this aspect of the experience, the *intentional quality* of the experience. Every intentional experience is also directed at something, is also about something, be it an experience of a deer, a cat, or a mathematic state of affairs. Husserl called the component that specifies what the experience is about, the *intentional matter* of the experience (Husserl 1984, 425–26). Needless to say, the same quality can be combined with different matters, and the same matter can be combined with different qualities. It is possible to doubt that 'the inflation will continue,' doubt that 'the election was fair,' or doubt that 'one's next book will be an international bestseller,' just as it is possible to deny that 'the lily is white,' to judge that 'the lily is white,' or to question whether 'the lily is white.' Husserl's distinction between the intentional matter and the intentional quality consequently bears a certain resemblance to the contemporary distinction between propositional content and propositional attitudes (though it is important to emphasize that Husserl by no means took all intentional experiences to be propositional in nature). But, and this is of course the central point, Husserl considered these cognitive differences to be *experiential* differences. Each of the different intentional qualities has its own phenomenal character. There is an *experiential* difference between affirming and denying that Hegel was the greatest of the German idealists, just as there is an *experiential* difference between expecting and doubting that Denmark will win the 2004 FIFA World Cup. What it is like to be in one type of intentional state differs from what it is like to be in another type of intentional state.[2] Similarly, the different intentional matters each have their own phenomenal character. There is an *experiential* difference between believing that 'thoughts without content are empty' and believing that 'intuitions without concepts are blind,' just as there is an *experiential* difference between denying that 'the Eiffel Tower is higher than the Empire State building' and denying that 'North Korea has a viable economy.' To put it differently, a change in the intentional matter will entail a change in what it is like to undergo the experience in question.[3] And these experiential differences, these differences in what it is like to think different thoughts, are not simply sensory differences.[4]

---

2 Using a decidedly Husserlian *jargon*, Siewert has recently spoken of *noetic phenomenal features* (Siewert 1998, 284). Sticking to the distinction between propositional content and attitude, one could argue that there is what one could call a qualitative feel to the different propositional attitudes.

3 However, this does not entail that two experiences that differ in their 'what it is like' cannot intend the same object, nor does it entail that two experiences that are alike in their 'what it is like' must necessarily intend the same object.

4 When we think a certain thought, for instance the thought 'Paris is the capital of France,' the thinking will often be accompanied by a non-vocalized utterance,

In the same work, Husserl also called attention to the fact that one and the same object can be given in a variety of different modes. This is not only the case for spatiotemporal objects (one and the same tree can be given from this or that perspective, as perceived or recollected, etc.), but also for ideal or categorial objects. There is an experiential difference between thinking of the theorem of Pythagoras in an empty and signitive manner, i.e., without really understanding it, and doing so in an intuitive and fulfilled manner, i.e., by actually thinking it through with comprehension (Husserl 1984, 73, 667–76). In fact, as Husserl points out, our understanding of signs and verbal expressions can illustrate these differences especially vividly: "Let us imagine that certain arabesques or figures have at first affected us merely aesthetically, and that we then suddenly realize that we are dealing with symbols

---

an aural imagery or auralization, of the very string of words used to express the thought. When we think the thought, we frequently 'hear' the sentence for our inner ear. At the same time, the thought will also frequently evoke certain 'mental images', say, visualizations of the Eiffel Tower, of baguettes, etc. It could be argued that abstract thoughts are accompanied by mental imagery and that the phenomenal qualities to be encountered in abstract thought are in fact constituted by this imagery. However, as Husserl already made clear in *Logical Investigations*, this attempt to deny that thinking has any distinct phenomenality to it is problematic. As Husserl points out, there is a marked difference between what it is like to auralize a certain string of meaningless noise, and what it is like to auralize the very same string, but this time understanding and meaning something by it (Husserl 1984, 46–47, 398). Since the phenomenality of the auralization is the same in both cases, the phenomenal difference must be located elsewhere, namely in the thinking itself. The case of homonyms and synonyms also clearly demonstrate that the phenomenality of thinking and the phenomenality of aural imagery can vary independently of each other. As for the attempt to identify the phenomenal quality of thought with the phenomenal quality of visualization a similar argument can be employed. Two different thoughts, say, 'Paris is the capital of France', and 'Parisians regularly consume baguettes,' might be accompanied by the same visualization of baguettes, but what it is like to think the two thoughts remain very different. Having demonstrated this much, Husserl then proceeds to criticize the view according to which the imagery actually constitutes the very meaning of the thought: To understand what is being thought is to have the appropriate 'mental image' before one's inner eye (Husserl 1984, 67–72). The arguments he employs bear a striking resemblance to some of the ideas that were subsequently used by Wittgenstein in *Philosophical Investigations*: 1. From time to time, the thoughts

or verbal signs. In what does this difference consist? Or let us take the case of a man attentively hearing some totally strange word as a sound-complex without even dreaming it is a word, and compare this with the case of the same man afterwards hearing the word, in the course of conversation, and now acquainted with its meaning, but not illustrating it intuitively. What in general is the surplus element distinguishing the understanding of a symbolically functioning expression from the uncomprehended verbal sound? What is the difference between simply looking at a concrete object *A*, and treating it as a representative of 'any *A* whatsoever'? In this and countless similar cases it is the act-characters that differ"(Husserl 1984, 398).

More recently, Galen Strawson has argued in a similar fashion, and in his book *Mental Reality* he provides the following neat example. Strawson asks us to consider a situation where Jacques (a monoglot Frenchman) and Jack (a monoglot Englishman) are both listening to the

---

we are thinking, for instance thoughts like 'every algebraic equation of uneven grade has at least one real root' will in fact not be accompanied by any imagery whatsoever. If the meaning were actually located in the 'mental images,' the thoughts in question would be meaningless, but this is not the case. 2. Frequently, our thoughts, for instance the thought that 'the horrors of World War I had a decisive impact on post-war painting' will in fact evoke certain visualizations, but visualizations of quite unrelated matters. To suggest that the meanings of the thoughts are to be located in such images is of course absurd. 3. Furthermore, the fact that the meaning of a thought can remain the same although the accompanying imagery varies also precludes any straight forward identification. 4. Absurd thoughts like the thought of a square circle is not meaningless, but it can never be accompanied by a matching image, since a visualization of a square circle is impossible in principle. 5. Finally, referring to Descartes' famous example in the *Meditations*, Husserl points out that we can easily distinguish thoughts like 'a chiliagon is a many-sided polygon,' and 'a myriagon is a many-sided polygon,' although the imagery that accompany both thoughts might be indistinguishable. Thus, as Husserl concludes, although imagery might function as an aid to the understanding, it is not what is understood, it does not constitute the meaning of the thought (Husserl 1984, 71). As Siewert later concludes (having independently covered much of the same ground as Husserl): Wittgenstein has "long warned us off the error assimilating thought and understanding to mental imagery. But we ought not to correct Humean confusion on this point, only to persist in the empiricist tradition's equally noxious error of supposing thought and understanding to be *experiential*, only if *imagistic*." (Siewert 1998, 305–6).

same news in French. Jacques and Jack are certainly not experiencing the same, for only Jacques is able to understand what is being said; only Jacques is in possession of what might be called an experience of understanding. To put it differently, there is normally something it is like, experientially, to understand a sentence. There is an experiential difference between hearing something that one does not understand, and hearing and understanding the very same sentence. And this experiential difference is not a sensory difference, but a cognitive one (Strawson 1994, 5–6). This is why Strawson then concludes as follows: "the apprehension and understanding of cognitive content, considered just as such and independently of any accompaniments in any of the sensory-modality-based modes of imagination or mental representation, is part of experience, part of the flesh or content of experience, and hence, trivially, part of the qualitative character of experience" (Strawson 1994, 12).[5]

Every conscious state, be it a perception, an emotion, a recollection, an abstract belief, etc., has a certain subjective character, a certain phenomenal quality, a certain quality of 'what it is like' to live through or undergo that state. This is what makes the mental state in question *conscious*. In fact, the reason we can be aware of our occurrent mental states (and distinguish them from one another) is exactly because there is something it is like to be in those states. The widespread view that only sensory and emotional states have phenomenal qualities must consequently be rejected. Such a view is not only simply wrong, phenomenologically speaking. Its attempt to reduce phenomenality to the "raw feel" of sensation marginalizes and trivializes phenomenal consciousness and is detrimental to a correct understanding of its cognitive significance.[6]

---

5 As Siewert has argued convincingly, the phenomenality of thinking is not a single phenomenally unvarying or monotonous experience. It is not as if the 'what it is like' to have conscious thoughts is the same, no matter what these thoughts are about; rather the phenomenal character of thinking is in continual modulation (Siewert 1998, 278–82).

6 For further attempts to argue in defence of a broader notion of phenomenal consciousness: cf. Smith (1989); Flanagan (1992); Goldman (1997); Van Gulick (1997); and Siewert (1998).

2. An intentionalistic interpretation of phenomenal qualities

It is one thing to argue that there is a phenomenal side to all conscious forms of intentionality, but what about the claim that all experiences have intentional features? Is that really true, or are there not a manifold of experiences which lack intentionality, say, feelings of pains or nausea, or moods like anxiety or nervousness?

The usual way to handle this problem in phenomenology has been by distinguishing two very different kinds of intentionality. In a narrow sense, intentionality is defined as object-directedness, in a broader sense, which covers what Husserl called operative [*fungierende*] intentionality, intentionality is defined as openness towards alterity and includes for instance our non-objectifying being-in-the-world. In both cases, the emphasis is on denying the attempt to understand consciousness as some kind of self-enclosed immanence.

If we go to the alleged non-intentional experiences with this distinction in mind, then it is true that pervasive moods such as sadness, boredom, nostalgia, or anxiety must be distinguished from intentional feelings like the desire for an apple or the admiration for a particular person. But although the moods in question are not types of object-intentionality, although they all do lack a specific intentional object, they are *not* without a reference to the world. They do not enclose us within ourselves but are lived through as pervasive affective atmospheres that deeply influence the way we encounter entities in the world. Just think, for example, of moods like curiosity, nervousness, or happiness. In fact, it has occasionally even been argued that moods, rather than being merely attendant phenomena, are rather fundamental forms of disclosure. We are always in some kind of mood. Even a neutral and distanced observation has its own tone, and as Heidegger famously wrote, "*Mood has always already disclosed being-in-the-world as a whole and first makes possible directing oneself toward something*" (Heidegger 1996, 129).

What about something like pain then? Well, Sartre's classical analysis in *Being and Nothingness* is, I think, illuminating. Assume that you are sitting late at night trying to finish a book. You have been reading most of the day and your eyes are hurting. How does this pain originally manifest itself? According to Sartre, not yet as a thematic object of reflection, but by influencing the way in which you perceive the world. You might become restless, irritated, have difficulties in focusing and concentrating. The words on the page may tremble or quiver. At this point, the pain is not yet apprehended as an intentional object, but that does not mean that it is either cognitively absent or unconscious. It is not yet reflected-upon as a psychic object, but rather given as a vision-in-pain, as an affective atmosphere that influences your intentional interaction with the world (Sartre 1956, 332–33).

The divide-and-rule strategy, the attempt to separate intentionality and phenomenality, the attempt to deny that intentional states have any intrinsic phenomenal properties, and that phenomenal states have any intrinsic intentional properties, and the attempt to treat each topic as if it could be understood in isolation from the other, leads not only very easily to a kind of "consciousness inessentialism," to the view that phenomenal consciousness is cognitively epiphenomenal. As mentioned earlier, the strategy also seems to reinstate a traditional concept of subjectivity that runs foul of everything that has been captured by the phrase 'being-in-the-world.' According to such a traditional (empiricist) concept, phenomenal consciousness has in and of itself no relation to the world. It is like a closed container filled with experiences that have no immediate bearing on the world outside. Typically, this internalist position has then been given a representationalist slant: On its own, our mind cannot reach all the way to the objects themselves. It is therefore necessary to introduce some kind of representational interface between the mind and the world if we are to understand and explain intentionality, i.e., the claim has been that our cognitive access to the world is mediated by mental representations.

In contrast, for the phenomenologists, subjectivity – the experiential dimension – is not a self-enclosed mental realm; rather, subjectivity and world are, as Merleau-Ponty puts it in his *Phenomenology of Perception*, co-dependent and inseparable (Merleau-Ponty 1962, 430). Subjectivity is essentially oriented and open toward that which it is not, and it is exactly in this openness that it reveals itself to itself. What is disclosed by the cogito is, consequently, not a self-contained immanence or a pure interior self-presence, but an openness toward alterity, a movement of exteriorization and perpetual self-transcendence.[7] Since the phenomenological theories of intentionality are unfailingly non-representationalist, they also reject the view according to which phenomenal experiences are to be conceived of as some kind of internal movie screen that confronts us with mental representations. We are 'zunächst und zumeist' directed at real existing objects, and this directedness is not mediated by any intra-mental objects. The so-called qualitative character of experience, the taste of a lemon, the smell of coffee, the coldness of an ice cube are not at all qualities belonging

---

7 This might sound like externalism. But actually, it is questionable whether the very choice between internalism and externalism, an alternative based on the division between inner and outer – is reference determined by factors *internal* to the mind, or by factors *external* to the mind? – is at all acceptable to the phenomenologists. Already in *Logical Investigations* Husserl argued that the notions of inner and outer, notions which he claimed expressed a naive commonsensical metaphysics, were inappropriate when it came to understanding the nature of intentionality (cf. Husserl 1984, 673, 708). This rejection of a commonsensical split between mind and world is even more pronounced after Husserl's transcendental turn. In *Cartesian Meditations*, for instance, Husserl writes that it is absurd to conceive of consciousness and true being as if they were merely externally related, when the truth is that they are essentially interdependent and united (Husserl 1976, 117. Cf. Husserl 1959, 432). If we pass on to Heidegger, he is also famous for having argued that the relation between Dasein and world could not be grasped with the help of the concepts 'inner' and 'outer.' As he writes in *Being and Time*: "In directing itself toward ... and in grasping something, Da-sein does not first go outside of the inner sphere in which it is initially encapsulated, but, rather, in its primary kind of being, it is always already 'outside' together with some being encountered in the world already discovered. Nor is any inner sphere abandoned when Da-sein dwells together with a being to be known and determines its character. Rather, even in this 'being outside' together with its object, Da-sein is 'inside' correctly understood; that is, it itself exists as the being-in-the-world which knows" (Heidegger 1996, 58). In my view, the phenomenological analyses of intentionality (be it Husserl's, Heidegger's, or Merleau-Ponty's) all entail such a fundamental rethinking of the very relation between subjectivity and environment that it no longer makes sense to designate them as being either internalist or externalist. This claim might be relatively uncontroversial when it comes to Heidegger and Merleau-Ponty, but it is controversial when it comes to Husserl, since he (at least by Anglo-American philosophers) is frequently

to some spurious mental objects, but qualities of the presented objects. Rather than saying that we experience *representations*, it would be better to say that our experiences are *presentational*, and that they *present* the world as having certain features.[8]

Reflections like these can also be found in analytical philosophy. Recently a number of analytical philosophers have criticized the view that phenomenal qualities are in and of themselves non-intentional and have instead defended what might be called an *intentionalistic* interpretation of phenomenal qualities. The point of departure has been the observation that it can often be quite difficult to distinguish a description of certain objects from a description of the experience of these very same objects. Back in 1903, G. E. Moore called attention to this fact and dubbed it the peculiar *diaphanous* quality of experience: When you try to focus your attention on the intrinsic features of experience, you always seem to end up attending to what the experience is *of*. And as Tye argues, the lesson of this transparency is that "*phenomenology ain't in the head*" (Tye 1995, 151). To discover what it is like, you need to look at what is being intentionally represented.[9] Thus, as the argument goes, experiences do not have intrinsic and non-intentional qualities of their own, rather the qualitative character of experience consists entirely, as Dretske writes, in the qualitative properties objects are experienced as having (Dretske 1995, 1). Or to put it differently, the phenomenal qualities are qualities of that which is represented. Differences in what it is like are actually intentional differences. Thus an experience of a red

---

interpreted as a prototypical internalist and methodological solipsist. However, I believe that this interpretation is based on something that approaches a complete misunderstanding of what Husserl is up to (including a misinterpretation of his concept of noema, and of his notion of phenomenological reduction), but it would lead too far to argue for this claim here. See however Zahavi (2003a).

8 This also happens to be Putnam's view (cf. Putnam 1999, 156). For a discussion of some of the many affinities between Putnam's recent reflections and views found in phenomenology, see. Zahavi (2003b).

9 In contrast to the phenomenologists, however, both Tye and Dretske are representationalists; they have no qualms speaking of experiences as representing the outside world.

apple is subjectively distinct from an experience of a yellow sunflower in virtue of the fact that different kinds of objects are represented. Experiences simply acquire their phenomenal character by representing the outside world. As a consequence, all phenomenal qualities are as such intentional. There are no non-intentional experiences. Thus, for Tye, pain (and I assume that he is only talking about physical pain) is nothing but a sensory representation of bodily damage or disorder (Tye 1995, 113).

Dretske's and Tye's intentionalistic interpretation of phenomenal qualities has the great advantage of staying clear of any kind of immanentism. As already mentioned, it also bears a certain resemblance to views found in phenomenology. This is in particular the case for the analysis offered by Sartre. Sartre is renowned for his very radical interpretation of intentionality. To affirm the intentionality of consciousness is, according to Sartre, to deny the existence of any kind of mental content (including any kind of sense-data or qualia) (Sartre 1956). There is nothing in consciousness, neither objects nor mental representations. It is completely empty. Thus, for Sartre, the being of intentional consciousness consists in its revelation of transcendent being. Sartre consequently takes the phenomenal qualities to be qualities of worldly objects, and certainly not to be located within consciousness. However, from the fact that consciousness is nothing apart from its revelation of transcendent being (or as Tye and Dretske would probably say, from the fact that it exhausts itself in its representation of external reality), Sartre would never infer that intentional consciousness is therefore no problem for reductionism. On the contrary, in his view, it is exactly the emptiness (or non-substantiality) of consciousness, that demonstrates its irreducibility.

Both Tye and Dretske explicitly criticize the attempt to draw a sharp distinction between the intentional or (re)presentational aspects of our mental lives and their phenomenal or subjective or felt aspects. But interestingly enough, their reason for attacking the separation is exactly the opposite of my own. By proposing an intentionalistic interpretation of phenomenality they hope to avoid the hard problem altogether. Why? Because if phenomenality is basically a question of intentionality, and if intentionality can be explained reductively in terms of functional or causal relations, one can accept the existence of phenomenality (neither Dretske nor Tye are eliminativists) and still remain a physicalist (Tye 1995, 153, 181).

I think this conclusion is wrong. The decisive difficulty for reductionism is not the existence of epiphenomenal qualia, qualia in the sense of atomic, irrelational, ineffable, incomparable, and incorrigible mental objects. And the hard problem does not disappear if one (rightfully) denies the existence of such entities, and if one, so to speak, relocates the phenomenal from the 'inside' to the 'outside.' The hard problem is not about the existence of non-physical *objects* of experience, but about the very existence of *subjective experience* itself, it is about the very fact that objects are *given* to us (cf. Rudd 1998).

When asked to exemplify the 'what it is like' quality of experience, one will often find references to what has traditionally been called secondary sense qualities, such as the smell of coffee, the colour of red silk, or the taste of a lemon. However, this answer reveals an ambiguity in the notion of 'what it is like.' Normally, the 'what it is like' aspect is taken to designate experiential properties. However, if our experiences are to have qualities of their own, they must be qualities over and above whatever qualities the intentional object has. But it is exactly the silk which is red, and not my perception of it. Likewise, it is the lemon that is bitter and not my experience of it. The *taste* of the lemon is a qualitative feature of the lemon and must be distinguished from whatever qualities my *tasting* of the lemon has. Even if there is no other way to gain access to the gustatory quality of the lemon than by tasting it, this will not turn the quality of the object into a quality of the experience. But in this case a certain problem arises. There is definitely something it is like to taste coffee, just as there is an experiential difference between tasting wine and water. However, when one asks for this quality and for this qualitative difference, it seems hard to point to anything beside the taste of coffee, wine, and water, though this is not what we are looking for. Should we consequently conclude that there is in fact nothing in the tasting of the lemon apart from the taste of the lemon itself?

However, this conclusion is overhasty, and it overlooks that there are two sides to the question of 'what it is like.' In *Ideas I*, Husserl distinguishes between the intentional object in 'the how of its determinations' (*im Wie seiner Bestimmtheiten*) and in 'the how of its givenness' (*im Wie seiner Gegebenheitsweisen*) (Husserl 1973,

303–4). Although this distinction is introduced as a distinction that falls within the noematic domain (rather than being a distinction between the noetic and the noematic domain), it nevertheless points us in the right direction: There is a difference between asking about the property the object is experienced as having (what does the object feel like to the perceiver) and asking about the property of the experience of the object (what does the perceiving feel like to the perceiver). Both questions pertain to the phenomenal dimension, but whereas the first question concerns a worldly property, the second concerns an experiential property.[10] Contrary to what both Dretske and Tye are claiming, we consequently need to distinguish between 1) what the object is like for the subject, and 2) what the experience of the object is like for the subject (cf. Carruthers 1998; McIntyre 1999).[11] However, insisting upon this distinction is not enough. The tricky part is to respect the lesson of transparency and to avoid misconstruing the experiential properties as if they belong to some kind of mental objects. It is not the case that worldly properties such as blue or sweet are matched one by one by experiential doublets of an ineffable nature (let us call them *blue or *sweet) and that both kind of properties are present in ordinary perception. So again, how then is the distinction to be cashed in phenomenologically?

We are never conscious of an object simpliciter, but always of the object as appearing in a certain way (as judged, seen, described, feared, remembered, smelled, anticipated, tasted, etc.). We cannot be conscious of an object (a tasted lemon, a smelt rose, a seen table, a touched piece of silk) unless we are aware of the experience through which this object is

10 To speak of worldly properties in this context should not be misunderstood. It does not entail any metaphysical claims concerning the subject-independent existence of the said properties. The claim being made is merely that the properties in question are properties of the experienced objects, and not of the experience of the objects.

11 Carruthers further argues that Dretske's and Tye's first-order representational theories of consciousness are incapable of accounting for the difference between these two aspects, and that a kind of higher-order representational theory is called for (Carruthers 1998, 209). I disagree with this view, but it would lead too far to present my criticism here. Cf. Zahavi (1999).

made to appear (the tasting, smelling, seeing, touching). But this is not to say that our access to, say, the lemon is *indirect*, namely mediated, contaminated, or blocked by our awareness of the experience, since the given experience is not itself an object on a par with the lemon, but instead constitutes the very access to the lemon. The object is given through the experience, and if there is no awareness of the experience, the object does not appear at all. If we lose consciousness, we (or more precisely *our* bodies) will remain causally connected to a number of different objects, but none of these objects will appear. In short, my experience of a red cherry or a bitter lemon is the way in which these objects are there for me. The red cherry is present for me, through my seeing it. I attend to the objects through the experiences. Experiences are not objects; rather they are accesses to objects. These accesses can take different forms; one and the same object (with the exact same worldly properties) can be given in a number of different modes of givenness, it can for instance be given as perceived, imagined, or recollected. Experiential properties are not properties like red or bitter; rather they are properties pertaining to these different types of access.

Although the different modes of givenness differ from one another, they also share certain features. One common feature is the quality of *mineness*, the fact that the experiences are characterized by a first-personal givenness. When I am aware of an occurrent pain, perception, or thought from the first-person perspective, the experience in question is given immediately, non-inferentially and non-criterially as *mine*.[12]

---

12 It could be objected that it is misleading to suggest that experiences can be given in more than one way. Either an experience is given from a first-person perspective, or it is not given at all. But I think this is a mistake. It is correct that experiences must always be given from a first-person perspective, otherwise they wouldn't be experiences, but this doesn't prevent them from being given from a second-person perspective as well. Let us assume that I crash, and that I am being scolded by the driver whose car I have just damaged. That the driver is angry is not something I infer on the basis of an argument from analogy, it is something I immediately experience (cf. Scheler 1973, 254; Merleau-Ponty 1964, 52–53). That I experience the anger of the other, doesn't imply that my experience is infallible (perhaps the driver is actually happy about the accident, since he can

Whereas the object of John's perception, along with all its properties, is intersubjectively accessible in the sense that it can in principle be given to others in the same way that it is given to John, John's perceptual experience itself is only given directly to John. Whereas John and Mary can both perceive the numerically identical same red cherry, each of them have their own distinct perception of it and can share these just as little as Mary can share John's bodily pain. Mary might certainly realize that John is in pain, she might even sympathize with John, but she cannot actually feel John's pain the same way John does. Mary has no access to the *first-personal givenness* of John's experience.

This first-personal givenness of experiential phenomena is not something quite incidental to their being, a mere varnish that the experiences could lack without ceasing to be experiences. On the contrary, it is this first-personal givenness that makes the experiences *subjective*. To put it differently, with a slightly risky phrasing, their first-personal givenness entails a built-in self-reference, a primitive experiential self-referentiality.[13]

In contrast to the redness of the tomato or the bitterness of the tea (both of which are worldly properties), the mineness characterizing the perception of the redness or bitterness is not a worldly property, but an experiential property. When asked to specify 'what the experience of the object is like for the subject,' this is exactly one of the features to mention. In short, the experiential dimension does not have to do with the existence of ineffable qualia, it has to do with this dimension of first-personal experiencing.

The 'what it is like' question has two sides to it: 'what is the object like for the subject' and 'what is the experience of the object like for the subject.' But although these two sides can be conceptually and

---

now finally get a new car, but he simply doesn't want to show his real feelings), nor that the anger of the other is given to me in the same way that it is given to the driver himself. The anger is exactly given from a second-person perspective to me. If one denies that experiences can be given in this way, if one, in other words, categorically denies that we can experience other's experiences, one is confronted with the threat of solipsism (cf. Zahavi 2001a).

13 For an extensive argumentation, cf. Zahavi (1999, 2002).

phenomenologically distinguished, distinguishability is not the same as separability. It is not as if the two sides or aspects of phenomenal experience can be detached and encountered in separation from each other. When I touch the cold surface of a refrigerator, is the sensation of coldness that I then feel a property of the experienced object, or rather a property of the experience of the object? The correct answer is that the sensory experience contains two different dimensions to it, namely a distinction between the *sensing* and the *sensed*, and that we can focus upon either. Phenomenology pays attention to the givenness of the object. But it does not simply focus on the object exactly as it is given; it also focuses on the subjective side of consciousness, thereby becoming aware of our subjective accomplishments and the intentionality that is at play in order for the object to appear as it does. When we investigate appearing objects, we also disclose ourselves as datives of manifestation, as those to whom objects appear.

To put it differently, when speaking of a first-person perspective, when speaking of a dimension of first-personal experiencing, it would be a mistake to argue that this is something that exclusively concerns the type of access that a given subject has to its own experiences, whereas the access to objects in the common world is independent of a first-person perspective, precisely in that it involves a third-person perspective. This line of thought will not do, and for the following reason: Obviously, I can be directed at intersubjectively accessible objects, but although my access to these objects is of the very same kind as the access of other persons, this does not imply that there is no first-person perspective involved. Rather, intersubjectively accessible objects are intersubjectively accessible precisely insofar as they can in principle be accessed directly from each and every first-person perspective. They thereby differ from experiences, which are in principle only directly accessible from the very same first-person perspective that they themselves constitute. Phrased differently, every givenness, be it the givenness of mental states, or the givenness of physical objects, involves a first-person perspective. There is no pure third-person perspective, just as there is no view from nowhere. To believe in the existence of such a pure third-person perspective is to succumb to an objectivist illusion. Of course, this is not to say that there is no third-person perspective, but merely that such a perspective is exactly a perspective from somewhere. It is a view that *we* can adopt on the world. It is a perspective that is founded upon a first-person perspective, or to be more exact, it emerges out of the encounter between at least two first-person perspectives, that is, it involves intersubjectivity.[14]

To summarize: The phenomenal dimension covers both domains: 1) what the object is like for the subject, and 2) what the experience of the object is like for the subject. The moment we are dealing with manifestation or appearance, we are faced with the phenomenal dimension. In fact, the 'what it is like' is exactly a question of how something appears to me, that is, it is a question of how it is given to and experienced by me. When I imagine a unicorn, desire an ice-cream, anticipate a holiday, or reflect upon an economic crisis, all of these experiences bring me into the presence of different intentional objects. What this means is not only that I am phenomenally acquainted with a series of worldly properties such as blue, sweet, or heavy; it also means that the object is there *for me* in different modes of givenness (as imagined, perceived, recollected, anticipated, etc). Both the worldly properties of the appearing object and the experiential properties of the modes of givenness are part of the phenomenal dimension. They are not to be separated, but neither are they to be confused.

In short, the wrong conclusion to draw from an intentionalistic interpretation of phenomenal qualities is that there is no hard problem of consciousness but only the easy problem of intentionality (information processing). The right conclusion to draw is that intentionality has a first-person aspect to it that makes it part of the hard problem and that it resists reductive explanation just as much as phenomenality does.

3. The Janus-face of experience

Even if it is true that intentionality and phenomenality are related, the nature of this relation still remains open for discussion. Is it intrinsic or extrinsic? Is it essential or merely contingent? To claim that it is

14 For a more elaborate argument, cf. Zahavi (2001b).

contingent, that is, to claim that intentionality is indifferent to whether it takes place in a conscious or unconscious medium is to subscribe to something McGinn has called the *Medium Conception*. According to this view, the relation between consciousness and intentionality is like the relation between a medium of representation and the message it conveys. On one side, we have the medium of sound, shape, or experience, and on the other, the content of meaning and reference. Each side can be investigated in separation from the other since their relation is completely contingent. Thus, according to this view, consciousness is nothing but a (rather mysterious) medium in which something relatively mundane, namely intentionality, is contingently embedded (McGinn 1991, 35). But is this really convincing?

On the face of it, what the experience is like and what it is of are by no means independent properties. Phenomenologists have typically argued that every appearance is an appearance of something for someone. Every appearance always has its *genitive* and *dative*. More recently, McGinn has made the same point, and has argued that experiences are *Janus-faced*: They have a world-directed aspect, they present the world in a certain way, but at the same time they also involve presence to the subject, and hence a subjective point of view. In short, they are of something other than the subject and they are like something for the subject, and as McGinn then continues: "But these two faces do not wear different expressions: for what the experience is like is a function of what it is of, and what it is of is a function of what it is like. Told that an experience is as of a scarlet sphere you know what it is like to have it; and if you know what it is like to have it, then you know how it represents things. The two faces are, as it were, locked together. The subjective and the semantic are chained to each other" (McGinn 1991, 29–30). In other words, the intentional and semantic content of an experience stands in an intimate relation to its phenomenal character and vice versa. But if what we are aware of is inextricably bound up with how it *appears* to us, phenomenal consciousness is not epiphenomenal, but rather cognitively indispensable.

Of course, it is possible to find a variety of different views on the matter. Some would say that intentionality is only a feature of conscious states, i.e., that only consciousness is in possession of genuine intentionality, and that any other ascription of intentionality is either derived or metaphorical (Searle 1998, 92–93). One line of argumentation in favour of such a view would be a line stressing the intrinsic connection between *experience*, *meaning*, and *intentionality*. As Strawson puts it: "[M]eaning is always a matter of something meaning something *to* something. In this sense, nothing means anything in an experienceless world. There is no possible meaning, hence no possible intention, hence no possible intentionality, on an experienceless planet.... There is no entity that means anything in this universe. There is no entity that is about anything. There is no semantic evaluability, no truth, no falsity. None of these properties are possessed by anything until experience begins. There is a clear and fundamental sense in which meaning, and hence intentionality, exists only in the conscious moment" (Strawson 1994, 208–9). Strawson consequently claims that experience is a necessary condition for genuine aboutness, and he suggests that there is an analogy between the sense in which a sleeping person might be said to be in possession of beliefs, preferences, etc., and the sense in which a CD might be said to contain music when it is not being played by a CD player. Considered merely as physical systems, neither of them are intrinsically about one thing rather than another, neither of them have any intrinsic (musical or mental) content. Strictly speaking, "it is no more true to say that there are states of the brain, or of Louis, that have intrinsic mental content, when Louis is in a dreamless and experienceless sleep, than it is true to say that there are states of a CD that have intrinsic musical content as it sits in its box" (Strawson 1994, 167).

However, apart from outright denying the existence of genuine non-conscious intentionality, there is also another option open. One might accept the existence of a non-conscious form of intentionality but still argue that non-conscious intentionality and conscious intentionality have nothing (or very little) in common, for which reason an elucidation of the first type of intentionality throws no light upon the kind of intentionality that we find in conscious life. It is not possible to account for the intentionality of my experience without accounting for the phenomenal aspect of the experience as well, and it is impossible to account for the phenomenal aspect of the experience without referring to its intentionality. Any discussion of intentional consciousness that left out the question of phenomenal consciousness (and vice versa) would be severely deficient. In short, when it comes to conscious intentionality we need an integrated approach. For this reason, Chalmers's distinction between the hard and the easy problems of consciousness is problematic. What he calls the easy problems of consciousness are either part of the hard problem or not about consciousness at all.

4. *Conclusion*

This article has had three aims: To problematize Chalmers's distinction between the hard and easy problems of consciousness; to offer some reflections on the relationship between intentionality and experience; and finally to point to some of the convergences between contemporary analytical philosophy of mind and phenomenology.

As for the first aim, I don't think anything needs to be added. As for the second, all that I have been able to offer have been some preliminary reflections, and there are of course many additional problems that have been left untouched. To mention a few: There is the question about the existence of the unconscious, and about so-called dispositional beliefs. How do they fit into the framework presented above?[15] There is also the entire discussion between internalism and externalism. Externalists

15 Does it make sense to speak of an unconscious tasting of coffee? An unconscious hearing and appreciation of Miles Davis? An unconscious desiring for chocolate? If an unconscious *experience* is to deserve its name, and not merely be an objective, physical process, it must presumably be subjective. After all, we do not call a stone, a table or the blood in our veins unconscious. But where is this subjectivity to manifest itself? Supposedly in the particular first-personal givenness of the experience. But it is difficult to imagine how an unconscious experience should possess such a feature. Unconscious experiences are per definition without a first-personal givenness; there is nothing it is like for the subject to have them. But can one really abstract the peculiar subjective givenness of the experience from the experience and still retain an experience, or is the ontology of experiencing not rather a first-person ontology? If it is a defining feature of an experience that there is necessarily something it is like for the subject to have it, it will be just as non-sensical to speak of an unconscious experience as to speak of an unconscious consciousness (something that even Freud refrained from doing, cf. Freud 1945, 434). Of course, this does

typically claim that differences in thought can be extraphenomenally fixed. If this is true, what implications does it have for the relation between intentionality and experience? Then there is the question of how an intentionalistic interpretation of phenomenal qualities can handle cases of hallucinations. And finally, what about the objection that the attempt to argue for an intimate relation between intentionality and experience implies some subtle form of psychologism? To claim that there is a special experience of understanding is bound to provoke the Wittgensteinians. But how should one defuse their criticism? All of these questions are topics in need of further treatment.

Let me spend some more time on the third issue, however. In recent years the issues of subjectivity, phenomenal consciousness, and selfhood have once again become central and respectable topics in analytical philosophy. This change in orientation has in general made analytical philosophers much more receptive to phenomenology. In fact, it is now almost commonplace to argue that any convincing theory of mind has to take phenomenology into account. However, this ready use of the term 'phenomenology' is to some extent rather misleading. When speaking of phenomenology, the vast majority of analytical philosophers are simply referring to a first-person description of what the 'what it is like' of experience is really like. Phenomenology is in other words identified with some kind of introspectionism. But for anybody familiar with

---

not exclude that there might be different non-conscious states and processes that play a causal role in our experience, but to speak of such non-conscious processes is not *per se* to speak of unconscious *experiences*. Flanagan has recently introduced a distinction between '*experiential sensitivity*' and '*informational sensitivity*.' Somebody may be experientially insensitive but informationally sensitive to a certain difference. When we aremerely informationally sensitive to something, we are not conscious of it, that is, pure informational sensitivity, or to use a better expression, pure informational pickup and processing is non-conscious. It is a processing without phenomenal awareness (Flanagan 1992, 55–56, 147). Subjectivity has to do with experiential sensitivity, and it is only the latter that lets us have phenomenal access to the object. But although it might be appropriate to operate with a notion of non-conscious informational processing, I think one should be careful not to assume that the informational sensitivity provides us with a non-phenomenal version of the exact *same* information as the experiential sensitivity. To suggest something like that is once again to flirt with the view that consciousness is cognitively epiphenomenal.

Continental philosophy, this notion of phenomenology will appear as both tame and lame. Given the recent developments in analytical philosophy of mind, it would make much more sense to engage in a discussion with the kind of phenomenology that was inaugurated by Husserl, and developed and transformed by, among many others, Scheler, Heidegger, Fink, Gurwitsch, Sartre, Merleau-Ponty, Lévinas, and Henry. The fact that subjectivity has always been of central concern to phenomenologists, and that they have devoted much time to a close scrutiny of the first-person perspective, the structures of experience, time-consciousness, body-awareness, self-awareness, intentionality, and so forth, makes them into obvious interlocutors.[16]

---

16 Back in 1995, in a book entitled *Approaches to Intentionality*, William Lyons gave a detailed overview of a number of contemporary theories of intentionality. He distinguished between an 'instrumentalist approach' (Dennett), a 'representationalist approach' (Fodor), a 'teleological approach' (Millikan), an 'information-processing approach' (Dretske), and a 'functionalist approach' (Loar). In the introduction prefacing his discussion of these different approaches, Lyons briefly remarked that contemporary theories deny the claim of Brentano and Husserl: That consciousness is essential to intentionality (Lyons 1995, 4). This appraisal provokes three critical questions: The first is whether Lyons account is already outdated. As part of the general 'consciousness boom,' the relation between consciousness and intentionality is currently once again up for discussion. The second is whether Lyons account was already outdated when he wrote it: Not only does it seem rather strange that Lyons ignored Searle's theory of intentionality, but even more to the point: Lyons apparently thought that theories of intentionality were an exclusive concern of analytical philosophy. Thus, he made no reference whatsoever to the theories of intentionality found in twentieth-century German and French thought. A final critical comment concerns Lyons's repeated tendency to place Brentano and Husserl side by side. Lyons is not alone in making them two of a kind (though it is rather strange to see Husserl's theory of intentionality described as nineteenth-century theory [Lyons 1995, 3]), but that Lyons's view is shared by many other analytical philosophers does not make it any more true. Not only are there absolutely crucial differences between Brentano's and Husserl's theory of intentionality (this is the case even for Husserl's early theory in *Logical Investigations*), but to have Brentano's theory of intentionality presented again and again as the sole Continental alternative to the approaches found in analytical philosophy, as is frequently being done by analytical philosophers, reveals an astonishing lack of familiarity with the philosophical tradition.

In my discussion of the relationship between intentionality and experience, I have tried to show that contemporary analytical philosophy of mind and (Continental) phenomenology have a number of common concerns.[17] However, in the beginning of the article I briefly mentioned that a more open exchange between the two traditions might not only help prevent further unintended repetitions, but that it might also bring the contemporary debate to a higher level of sophistication. I am not sure that I have already succeeded in demonstrating the later claim. This is in part because I chose to focus on a number of existing similarities and overlaps. This choice was propaedeutically motivated: If an exchange is to be encouraged, it is first necessary to show that the much-discussed gap between analytical philosophy and continental philosophy, a gap which has often been taken to be so wide that it prevents any kind of dialogue, is a fiction. However, although the mere existence of overlaps might be fascinating, if the discussion is to move forward it is not the overlaps that are of real interest, it is the (relevant) *differences*. In other words, the reason why analytical philosophers should pay more attention to phenomenology is not because the latter tradition contains analyses that are fully up to date with what is currently going on in analytical philosophy of mind. No, the real reason is of course that phenomenology is still way ahead of analytical

17 Another area is the issue of self-awareness. The phenomenological investigation of self-awareness has typically been set in the context of a discussion of such diverse issues as spatiality, embodiment, temporality, intersubjectivity, attention, and so forth. However, it is also a characteristic feature of recent analytical philosophy that an increasing number of philosophers have distanced themselves from traditional armchair philosophy and abandoned the attempt to capture the basic structures of mind solely by means of a priori conceptual analysis. Instead, they have started to engage in dialogue with empirical science, and to draw upon the resources found in cognitive science, psychopathology, neuropsychology, and developmental psychology. As a result, they have become aware of the interplay between subjectivity, embodiment, and environment, and have reached conclusions on issues such as the existence of prelinguistic forms of self-awareness, the bodily roots of self-experience, and the connection between exteroception and proprioception, that all bear a striking resemblance to views already found in phenomenology. Cf. Zahavi (1999, 2002).

philosophy when it comes to the investigation of certain aspects of consciousness. In contrast to many analytical philosophers, the phenomenologists have never taken the problem of consciousness to be first and foremost a question of how to relate consciousness to the brain, first and foremost a question of how to reduce consciousness to mind- and meaningless matter. Not only have they considered this enterprise to be futile for various conceptual reasons, but they have also typically argued that such a take completely overlooks the urgent need for a thorough investigation of phenomenal consciousness on its own terms. To put it differently, although the phenomenologists have typically conceived of the experiential dimension as being so fundamental that no non-circular explanation of it is possible, they would deny that such an outlook puts a hold on further analysis. After all, there is much more to the question of phenomenal consciousness than a mere recognition of its irreducibility. A thorough elucidation of its structures requires a closer investigation of such issues as selfhood, first-personal givenness, attention, thematic and marginal consciousness, reflective and pre-reflective self-consciousness, inner time-consciousness, body-awareness, etc.

Given the richness of the phenomenological analyses of consciousness, and given that many of the conclusions that have lately been reached by analytical philosophers are in fact rediscoveries, the habitual stance of analytical philosophy towards phenomenology – which has ranged from complete disregard to outright hostility – can only be characterized as counterproductive. But just as phenomenology has something to offer analytical philosophy, phenomenology can certainly also profit, not only from the analytical discussions of, for instance, indexicality, the first-person perspective, internalism vs. externalism, and the possibility of pre-linguistic experience, but also from the conceptual clarity and problem-oriented approach found in analytical philosophy. Thus, the very attempt to engage in dialogue with analytical philosophy might hopefully force phenomenology to become more problem-oriented and thereby counteract what is currently one of its greatest weaknesses: its preoccupation with exegesis. One concrete step would be for those trained in phenomenology to make more of an attempt to formulate their reflections in a relatively non-technical manner. Much could be achieved by such a gesture. It would be bound to facilitate constructive discussions with those figures in analytical philosophy that more or less on their own have started to work on phenomenological themes. And it would be a pity to miss the opportunity for dialogue that is currently at hand.[18]

18 I am indebted to James Hart, Eduard Marbach, John Drummond, and Anthony Steinbock for comments to an earlier version of this article. The study was funded by the Danish National Research Foundation.

References

Carruthers, P. (1998). "Natural Theories of Consciousness," *European Journal of Philosophy* **6**: 203–22.

Chalmers, D. J. (1995). "Facing Up to the Problem of Consciousness," *Journal of Consciousness Studies* **2(3)**: 200–19.

Chalmers, D. J. (1996). *The Conscious Mind. In Search of a Fundamental Theory*. New York: Oxford University Press.

Chalmers, D. J. (1997). "Moving Forward on the Problem of Consciousness," *Journal of Consciousness Studies* **4(1)**: 3–46.

Dretske, F. (1995). *Naturalizing the Mind*. Cambridge, MA: MIT Press.

Flanagan, O. (1992). *Consciousness Reconsidered*. Cambridge, MA: MIT Press.

Freud, S. (1945). *Gesammelte Werke VIII. Werke aus den Jahren 1909–1913*. S. Fischer: Frankfurt am Main.

Goldman, A. I. (1997). "Consciousness, Folk Psychology, and Cognitive Science." In *The Nature of Consciousness*, ed. N. Block, O. Flanagan, and G. Güzeldere. Cambridge, MA: MIT Press, 111–25.

Heidegger, M. (1996). *Being and Time*. Albany: SUNY Press.

Hodgson, D. (1996). "The Easy Problems Ain't So Easy," *Journal of Consciousness Studies* **3(1)**: 69–75.

Husserl, E. (1959). *Erste Philosophie II (1923–24)* [*Husserliana* VIII]. Den Haag: M. Nijhoff.

Husserl, E. (1973). *Cartesianische Meditationen und Pariser Vorträge* [*Husserliana* I]. Den Haag: M. Nijhoff.

Husserl, E. (1976). *Ideen zu einer reinen Phänomenologie und phänomenologischen Philosophie I* [*Husserliana* III/1–2]. Den Haag: M. Nijhoff.

Husserl, E. (1984). *Logische Untersuchungen II* [*Husserliana* XIX/1–2]. Den Haag: M. Nijhoff.

Jacob, P. (1998). "What Is the Phenomenology of Thought?" *Philosophy and Phenomenological Research* **58**: 443–48.

Lyons, W. (1995). *Approaches to Intentionality*. Oxford: Clarendon Press.

McGinn, C. (1991). *The Problem of Consciousness: Essays Toward a Resolution*. Oxford: Blackwell.

McIntyre, R. (1999). "Naturalizing Phenomenology? Dretske on Qualia." In *Naturalizing Phenomenology*, ed. J. Petitot, F. J. Varlea, B. Pachoud, and J.-M. Roy. Stanford: Stanford University Press, 429–39.

Merleau-Ponty, M. (1962). *Phenomenology of Perception*. Trans. Colin Smith. London: Routledge Press.

Merleau-Ponty, M. (1964). *Sense and Non-Sense*. Evanston, IL: Northwestern University Press.

Moore, G. E. (1903). "The Refutation of Idealism," *Mind* **12**: 433–53.

O'Shaughnessy, B. (2000). *Consciousness and the World*. Oxford: Oxford University Press.

Putnam, H. (1999). *The Threefold Cord. Mind, Body, and World*. New York: Columbia University Press.

Rudd, A. J. (1998). "What It's Like and What's Really Wrong with Physicalism: A Wittgensteinian Perspective," *Journal of Consciousness Studies* **5(4)**: 454–63.

Sartre, J.-P. (1956). *Being and Nothingness*. Trans. Hazel Barnes. New York: Philosophical Library.

Scheler, M. (1973). *Wesen und Formen der Sympathie*. Bern / München: Francke Verlag.

Searle, J. R. (1998). *Mind, Language and Society*. New York: Basic Books.

Siewert, C. P. (1998). *The Significance of Consciousness*. Princeton: Princeton University Press.

Smith, D. W. (1989). *The Circle of Acquaintance*. Dordrecht: Kluwer.

Strawson, G. (1994). *Mental Reality*. Cambridge, MA: MIT Press.

Tye, M. (1995). *Ten Problems of Consciousness*. Cambridge, MA: MIT Press.

Van Gulick, R. (1997). "Understanding the Phenomenal Mind: Are We All Just Armadillos?" In N. Block, O. Flanagan, and G. Güzeldere, eds., *The Nature of Consciousness*, Cambridge, MA: MIT Press, 559–66.

Zahavi, D. (1999). *Self-awareness and Alterity. A Phenomenological Investigation*. Evanston: Northwestern University Press.

Zahavi, D. (2001a). "Beyond Empathy: Phenomenological Approaches to Intersubjectivity," *Journal of Consciousness Studies* **8(5–7)**: 151–67.

Zahavi, D. (2001b). *Husserl and Transcendental Intersubjectivity*. Athens, OH: Ohio University Press.

Zahavi, D. (2002). "Phenomenology of the Self." In A. David and T. Kircher, eds., *The Self and Schizophrenia. A Neuropsychological Perspective*, Cambridge: Cambridge University Press.

Zahavi, D. (2003a). *Husserl's Phenomenology*. Stanford: Stanford University Press.

Zahavi, D. (2003b). "Natural Realism, Anti-reductionism, and Intentionality: The 'Phenomenology' of Hilary Putnam." In David Carr and Chan Fai Cheung, eds., *Time, Space and Culture*. Dordrecht: Kluwer.

CANADIAN JOURNAL OF PHILOSOPHY
Supplementary Volume 29

Redrawing the Map and Resetting the Time: Phenomenology and the Cognitive Sciences[1]

SHAUN GALLAGHER AND FRANCISCO J. VARELA[2]

In recent years there has been some hard-won but still limited agreement that phenomenology can be of central and positive importance to the cognitive sciences. This realization comes in the wake of dismissive gestures made by philosophers of mind who mistakenly associate phenomenological method with untrained psychological introspection (e.g., Dennett 1991). For very different reasons, resistance is also found on the phenomenological side of this issue. There are many thinkers well versed in the Husserlian tradition who are not willing to consider the validity of a naturalistic science of mind. For them cognitive science is too computational or too reductionistic to be seriously considered as capable of explaining experience or consciousness.[3] In some cases, when phenomenologists have seriously engaged the project of the cognitive sciences, rather than pursing a positive rapprochement with this project, they have been satisfied in drawing critical lines that identify its limitations.

---

1 Originally published in electronic form in *The Reach of Reflection: The Future of Phenomenology*, ed. S. Crowell, L. Embree and S. J. Julian (17–45). Electron Press (at http://www.electronpress.com/reach.asp).

2 Francisco Varela died on May 28, 2001.

3 Signals of this kind of suspicion have been sent recently by Paul Ricoeur in his conversation with Jean-Pierre Changeux (Changeux and Ricoeur 2000). Ricoeur suggests that phenomenology stands opposed to the cognitive sciences.

On the one hand, such negative attitudes are understandable from the perspective of the Husserlian rejection of naturalism, or from strong emphasis on the transcendental current in phenomenology. On the other hand, it is possible to challenge these attitudes from perspectives similar to the one taken by Merleau-Ponty (1962, 1964), who integrated phenomenological analyses with considerations drawn from the empirical sciences of psychology and neurology long before cognitive science was defined as such. In constructing a cross-disciplinary tradition like the one Merleau-Ponty inspires, thinkers in the relevant disciplines have to wrestle with a variety of issues, including the issue of naturalism. In this respect, however, natural scientists, more readily than phenomenologists, have come to acknowledge that phenomenology is directly relevant for a natural scientific understanding of cognition (e.g., Varela 1996; Varela et al. 1991). Even the hardest of the hard scientists have made recent peace offerings to phenomenology. For example, the neuroscientist Jean-Pierre Changeux, in his conversation with Paul Ricoeur, declares that his purpose "is not to go to war against phenomenology; to the contrary, [he wants] to see what constructive contribution it can make to our knowledge of the psyche, acting in concert with the neurosciences" (Changeux and Ricoeur 2000, 85). And Alain Berthoz, a neuroscientist who studies motor and perceptual systems, does not hesitate to invoke Husserl's analysis of time-consciousness in his explication of anticipatory aspects of motor control (Berthoz 2000, 16).[4]

In this chapter we explore the various ways in which phenomenology and the cognitive sciences can come together in a positive and productive exchange. In the first part, after some brief remarks about the nature of the cognitive sciences and the problem of naturalization, we begin by mapping out several issues that would benefit from this exchange. In the second part of the chapter we ask if this cross-disciplinary approach can address one of the most basic problems defined by

4 Both Changeux and Berthoz are at the College de France, where Merleau-Ponty has seemingly had a lasting influence.

Husserlian phenomenology, the problem of time-consciousness, and whether Husserl's analysis of this theme has anything to contribute to the cognitive sciences.

Part I: Defining the Issues

A Different Cognitive Science and a Different Phenomenology

If one begins by thinking of cognitive science as it was first formulated in opposition to behaviourism, in terms of computational analysis and information processing, it is difficult to see how phenomenology might participate in the "Cognitive Revolution." On this formulation, the scientific study of cognition is a study of how the subpersonal, non-phenomenological mind manipulates discrete symbols according to a set of syntactical procedures, and how this might be cashed out in neurological terms. This, however, is no longer the current view of cognitive science. Faced with a variety of problems implicit in this view, the cognitive revolution took a different turn in the late 1980s. This corresponded to a new emphasis on neuroscience, and connectionism, which challenged the prevailing computational orthodoxy by introducing an approach based on nonlinear dynamical systems (see, e.g., Port and van Gelder 1995). With this formulation there was a shift away from an emphasis on reductionism to an emphasis on the notions of emergence and self-organization. The question was how higher-level personal structures emerged from lower-level subpersonal, self-organizing processes.

This turn in the fortunes of cognitive science also motivated a new interest in consciousness. It is nothing short of ironic that just when many phenomenologists were trading in their volumes of Husserl and Sartre for the texts of poststructural analysis, and thus abandoning the very notion of consciousness, philosophers of mind, who had started their work on ground circumscribed by Ryle's behaviouristic denial of consciousness, were beginning to explore the territory left behind by the phenomenologists. And while this continental drift away from consciousness was motivating a remapping of philosophical interests, theorists in the cognitive sciences were working their way toward a necessary rendezvous with phenomenology.

The current situation in the cognitive sciences is characterized by a growing interest in the ecological-embodied-enactive approach (Bermudez et al. 1995; Clark 1997; Varela et al. 1991). This approach takes up the connectionist emphasis on dynamical mechanisms and self-organizing emergence, but it further insists that cognition is best characterized as belonging to embodied, situated agents – agents who are *in-the-world*. On this understanding of the cognitive sciences, just as neuroscientists and neuropsychologists work together with researchers in artificial intelligence and robotics, so also phenomenologists and philosophers of mind work together with the empirical scientists in order to develop a fuller and more holistic view of cognitive life – a life that is not just the life of the mind, but of an embodied, ecologically situated, enactive agent.

This recent redefinition of the cognitive sciences, if it is to include a place for phenomenology, requires that we also conceive of phenomenology in a different way. Or at least we need to see that there is a section of the phenomenological map that can be redrawn along lines that reach across the theoretical divides that separate phenomenology from the sciences. One way to think of this is to think of *naturalizing phenomenology*. For many phenomenologists, this will seem self-contradictory, an antilogy. Phenomenology just is, by definition, non-naturalistic. For many others, the difficult question is how it might be accomplished without it losing the specificity of phenomenology. Everything, however, depends on what one means by naturalization. There is no question of considering here all possible proposals in this regard (see Roy et al. 1999, for a more detailed analysis). But consider two among many.

(1) The subjective data developed in phenomenology should be made objective, and thus amenable to scientific analysis. This suggestion is similar to Nagel's (1974) idea of an "objective phenomenology" that would allow for a level of abstraction from the particularism of individual reports, or to Dennett's (1991) idea of a "heterophenomenology" that would treat phenomenological reports as part of the objective data of science.

(2) Naturalization in the minimal sense means "not being committed to a dualistic kind of ontology" (Roy et al. 1999, 19). This includes the idea that phenomenology has to be explanatory and not just descriptive. Phenomenology would help to resolve the "explanatory gap," and contribute to an explanation of how brain and bodily processes can give rise to phenomenological properties that are not non-physical properties.

The first proposal is problematic in several ways. It equates phenomenology with folk psychology and understands phenomenological data to mean anything that a subject happens to report. In effect, because there is no concern about phenomenological method, this kind of approach, whether it follows a reductionist strategy or pursues the intentional stance, fails to take phenomenology seriously. Naturalization means, in this case, getting rid of phenomenology. Furthermore, if in such naturalizing strategies one does not employ phenomenological method, the very objectivity that is sought will be seriously compromised. For example, in translating a subject's first-person experience into third-person data, it would not do for the scientist to rely on her own subjective experience as an interpretational guide, since this would simply lead to a pollution of the subject's first-person data with the scientist's first-person data. A scientist would have to resort to objectively formalized meanings established within the framework of behavioural science in order to properly interpret the subject's reports. In that case, however, one needs to ask where these formalized meanings (generalizations and abstractions) originate. One comes quickly to the realization that at some point a controlled form of phenomenological experience is required to justify the formalized meanings used as interpretational guides. In effect, an objective interpretational framework depends upon a reflective, methodically guided phenomenological analysis, without which the procedure may simply impose the results of previous uncontrolled and anonymous phenomenological exercises (see Gallagher 1997).

The second proposal, as it is worked out by Roy et al. (1999), requires a phenomenological practice guided by method. What allows Husserlian phenomenology to escape from a naturalistic framework (the natural attitude) is a change of attitude achieved through a methodical practice (the phenomenological reduction). To move in the opposite direction, that is, to bring phenomenology to bear on the naturalistic enterprise of the cognitive sciences, involves another change of attitude. This does not mean abandoning phenomenological methods, but taking what we learn about first-person experience within the phenomenological attitude and using it in the context of naturalistic explanation. Although Husserl defined phenomenology as a non-naturalistic discipline, the idea that the results of his transcendental science might inform the natural sciences is not inconsistent with his own intent. He suggested, quite clearly, that "every analysis or theory of transcendental phenomenology – including ... the theory of the transcendental constitution of an objective world – can be developed in the natural realm, by giving up the transcendental attitude" (1970, §57).

Roy et al. (1999) develop one possible program that would lead to a naturalization of phenomenology in this sense. The issue is how to transpose the results from phenomenological, methodical analysis to the natural context, without stopping at statements of mere correlation. They propose a re-categorization of phenomena at a level of abstraction sufficient to allow for the recognition of common properties between phenomenological data and objective data developed in the sciences. The task would involve moving to a level of explanation that would be abstract enough to embrace all of the data. One possibility would involve a mathematical interpretation, a transformation of concepts into algorithms similar to transformations of this kind found in the physical sciences. In effect, if one could develop a formal language to express phenomenological findings, using perhaps a phenomenological notation of the sort suggested by Marbach (1993), the task would be to integrate it with a similar formal expression of physical processes. The appeal to mathematics is an appeal to formalized and intersubjectively verifiable meanings within a common language.

Another possible route to naturalization is to view phenomenology and the cognitive sciences as mutually constraining (Varela 1996; Gallagher 1997). For example, there are explanations of voluntary action that are developed at the level of cognitive mechanisms. A cognitive mechanism is usually thought of as a heuristic that will eventually be cashed out in terms of neurological processes. Can such an explanation succeed if it fails to make sense out of the rich phenomenological experience that accompanies voluntary action? This does not mean that cognitive explanations have to identify physical processes that are isomorphic with phenomenological data; but at a minimum, if a cognitive explanation implies or requires a phenomenological correlation that is unlikely or impossible, some negotiation has to take place between the two levels of description. And this does not necessarily mean that phenomenology will either win or lose. It is quite possible that the mutual constraint situation will lead to a productive mutual enlightenment, where progress in the cognitive sciences will motivate a more finely detailed phenomenological description developed under the regime of phenomenological reduction, and a more detailed phenomenology will contribute to defining an empirical research program. A specific example concerning preparation strategies and visual perception has been provided recently (Lutz et al. 2002). In this study phenomenological reduction was used as the basis for selecting electrical events. This, in turn, validated the various phenomenological clusters of reports provided by the subjects.

Although we have invoked phenomenological reduction as crucial, it is also true that there is little agreement and even less explicit development of the *pragmatics* that should define its role in any non-reductionist naturalization project. In other words, explicit accounts of how to carry out the gestures involved in the *epoché* have to be developed, and how these are to be linked to intersubjective validation. This is a central issue, but it will not concern us here (for recent attempts towards a disciplined phenomenological pragmatics see Depraz et al. 2000; 2003).

Back to the Issues Themselves

Consider some of the problems that concern phenomenologists. How do I perceive space? How is perception different from memory, or dream-consciousness, or fantasy? When I remember or imagine something, does my thinking consist of images, or is my thinking more sentential? Does consciousness have a formal structure independently of its contents? In normal voluntary movement, to what extent, and in what sense modalities, am I aware of my body? As I move through my immediate environment, of what am I aware? How do I understand what another person is thinking?

Although philosophers of mind work in a very different tradition, they are interested in the same kinds of problems. Not unlike Husserl, many of them will attempt to resolve these issues by appealing directly to experience. Others, not unlike Merleau-Ponty, will consider these issues in the light of what empirical studies have to say about them. Indeed, it is equally the case that certain empirical scientists are interested in exactly the same issues. Moreover, all of these groups share a common goal. They want to attend to the issues themselves; they want to understand the basics of human experience. What is it that keeps us from searching together? Perhaps the kinds of answers that we seek are different. Nonetheless, it seems clear that a more complete understanding of these things can be developed if we can see how the various answers line up with each other.

It will be impossible to provide here a complete inventory of all the many issues that are of common concern among phenomenologists, philosophers of mind, and cognitive scientists. But we can suggest, in regard to a few of the issues, how phenomenology can contribute to and learn from the other disciplines. To that end, we will discuss two sets of issues that are central to the concerns of phenomenologists and cognitive scientists: embodied self-consciousness and intersubjectivity.

Embodied Self-Consciousness

Under the title of self-consciousness there are a large number of specific problems. Even if we focus on a minimal sense of embodied self-consciousness, the issue is complex. Distinctions that are familiar to phenomenologists, such as those between objective body (*Körper*) and lived body (*Leib*), are not always acknowledged in the cognitive sciences. Yet very detailed discussions concerning the nature of proprioception and body schemas, found in the cognitive sciences, can enrich the phenomenological distinction. Clarifications can be introduced on both sides. For example, a distinction between body image and body schema, which is often lost within phenomenological discussions, can be made clear by considering pathological cases involving neurological damage and loss of proprioception. Such cases can motivate a more detailed phenomenological analysis which can, in turn, contribute to further clarifications of the empirical cases (e.g., Gallagher and Cole 1995). The outcome of such clarifications informs philosophical discussions of primary (non-conceptual, pre-reflective) self-consciousness as it develops in the bodily experience of early infancy. Such themes were important to Merleau-Ponty (1962, 1964) and continue to be studied today by both philosophers and developmental psychologists (Gallagher and Meltzoff 1996).

Other related distinctions call for phenomenological clarification. For example, discussions in philosophy of mind often focus on the concept of the sense of ownership for movement, action, and thought (see, for example, Campbell 1999a). One finds, however, that the concept of ownership is complex. It involves distinguishing between senses of ownership for one's own body, for movement, and for action. Action also involves the notion of a sense of agency, which can be distinguished from the sense of ownership. Yet one does not find this distinction carefully made in the philosophical discussions. A phenomenological analysis of the difference between active and passive movement helps to make the distinction.

One would begin this analysis by bracketing all scientific theories about movement and motor control and by attending directly to one's own experience. If we define the *sense of agency* as the sense that I am the one who is causing or generating an action, and the *sense of ownership* as the sense that I am the one who is undergoing an experience, then in the phenomenology of voluntary or willed action, these two senses seem to be indistinguishable. When I intentionally reach for a cup and grasp it, I know this to be my own action. Agency coincides with ownership. This coincidence may be what leads philosophers to think of ownership of action in terms of agency: that the owner of an action is the person who is, in a particular way, causally involved in the production of that action. In the case of *involuntary* action, however, it is quite possible to distinguish, phenomenologically, between sense of agency and sense of ownership. I may have a sense that I am the one who is moving or is being moved, and thereby acknowledge ownership of the movement. I can self-ascribe it as *my* movement. At the same time, I may not have a sense of causing or controlling the movement, that is, no sense of agency. The agent of the movement is someone else – the person who pushed me from behind, the physician who is manipulating my arm in a medical examination, etc. My claim of ownership (my self-ascription that I am the one who is undergoing such experiences) is perfectly consistent with my lack of a sense of agency (Gallagher 2000).

Working out a simple phenomenological distinction, however, cannot be the end of the story. If the distinction is going to do anything for philosophers in their analysis of action, or for scientists in their analysis of motor control, one needs to carry this distinction back into the empirical discussions. For example, can one find empirical studies that confirm the distinction? If one can, it provides empirical confirmation of the phenomenological analysis, and at the same time, it offers a clear distinction to guide further scientific research. In the scientific literature one can indeed find evidence that supports the phenomenological distinction. Experiments involving certain pathological cases demonstrate a clear dissociation between two different systems of motor control:

(1) a sensory-feedback mechanism that compares intended movement with actual movement, by means of visual and proprioceptive feedback; and
(2) a "forward," pre-action mechanism that compares motor intention with motor commands (see, e.g., Fourneret and Jeannerod 1998; Frith and Done 1988).

In some pathological cases the failure of the forward mechanism corresponds to a lack of a sense of agency. For example, a schizophrenic subject who suffers delusions of control complains that *his* hand is moving (that is, he has a sense of ownership for the movement) but that *he* is not moving it (that is, no sense of agency). Such subjects in experimental circumstances are able to control their movement through sensory-feedback but are unable to control it through the quicker forward mechanism (Frith and Done 1988).

What seems to be a correlation between the phenomenological distinction (sense of agency and sense of ownership for movement) and the neurological distinction (forward control mechanism and sensory-feedback mechanism) calls for further research (see, for example, de Vignemont 2000; Franck et al. 2001). If the correlation holds up, not only will it suggest a neurological basis for these two aspects of bodily self-consciousness, but it will also provide a scientifically justified distinction that will clarify a variety of philosophical discussions that are in great need of clarification.

Knowledge of Others

An equally complex issue has a long history in the phenomenological tradition. We find the question of intersubjectivity central to the thoughts of Scheler, Husserl, Heidegger, Sartre, Merleau-Ponty, Ricoeur, and Levinas, to name a few of the thinkers who have wrestled with this problem. Again, there is an equally rich discussion ongoing in the philosophy of mind and the cognitive sciences. Moreover, there are some interesting parallels to be found in the two discussions. For example, criticisms of the notion of analogical inference – that is, that I understand the other as a mind only by an analogy I draw between her bodily behaviour and my own – have been outlined in both traditions. Both traditions are also marked by competing theories and the lack of consensus. On the phenomenological side, there are proponents of empathy and those who go beyond empathy (see Zahavi 2001, for a good review). On the cognitive side, there are those who favour "theory theory" (e.g., Baron-Cohen 1995; Leslie 2000), and those who champion the simulation approach (e.g., Gordon 1986, 1995; Goldman 1989). Rarely, however, does one find any crossover between the insights developed in the phenomenological discussions and the theories that make up the cognitive approach. Ironically, it is as if one discussion of knowing the other does not know about the other discussion (but see the essays in Thompson 2001 for an excellent exception to this problem).

Many of the claims made on the cognitive side, under the heading of "theory of mind," are based on neurological studies and experiments performed in the context of developmental psychology. The interpretation of these scientific studies, however, could benefit greatly from the guidance of phenomenology. Many of the assumptions made in the cognitive interpretations do not hold up to phenomenological analysis. For example, if we are to accept the cognitive explanations (whether they are provided by theory theorists or simulation theorists), then we would have to accept the idea that our primary interactions with others involve attempts to *explain* their mental states or to *predict* their behaviours. For example, theory theorists cite compelling evidence from false belief experiments with young children (e.g., Wimmer and Perner 1983). The experiments show that children at three years of age are unable to offer explanations or predictions that would distinguish their own point of view from the perspective of others. At ages four and above, however, most children are able to understand and explain false beliefs in others and are thus able to predict the other's behaviour. These studies are interpreted (and often designed) to support the idea that children develop a theory of mind – that is, a theory about the mental states and behaviours of others – and at a certain point in development are able to use this theory to explain the other's beliefs and, on that basis, predict their behaviour.

In many authors this idea is then generalized to the claim that, once formed, we use theory of mind as our primary means for understanding other persons (e.g., Changeux in Changeux and Ricoeur 2000, 154–57; Tooby and Cosmides 1995). For some theorists, the mentalistic framework seems to be the only possible framework for understanding others, and they consider it "our natural way of understanding the social environment" (e.g., Baron-Cohen 1995, 3–4).

Can phenomenologists, who have themselves struggled with the assignment of primacy to mentalistic frameworks, offer any insights that might be useful in this context? We think there is good phenomenological evidence to the effect that, in the majority of intersubjective situations, the human subject does not normally posit a theoretical entity, called a mental state, and then attribute it to the other person (this has been a long-standing line of thought in phenomenology; see e.g. Depraz 1995; Gallagher 2001). We do not interact with the other by conceiving of her mind as a set of *cogitationes* closed up in immanence (Merleau-Ponty 1962, 353). Both theory theory and simulation theory conceive of communicative interaction between two people as a process that takes place between two Cartesian minds. Children with a theory of mind supposedly "see people as living their lives within a world of mental content that determines how they behave in the world of real objects and acts." They construe "people's real-world actions as *inevitably* filtered through representations of the world rather than linked to the world directly" (Wellman 1993, 31–32). This assumes that one's understanding involves a retreat into a realm of *theoria* or *simulacra,* into a set of internal mental operations that come to be expressed (externalized) in speech, gesture, or interaction. If, in contrast, along with Merleau-Ponty, we think of communicative interaction as being accomplished in the very action of communication, in the speech, gesture, and interaction itself, then the idea that the understanding of another person involves an attempt to theorize about an unseen mental state, or to "mind-read," is problematic.

This is not to say that one cannot find the proper resources within the cognitive sciences that would correct the mentalistic approach of theory of mind. A large body of research into "primary intersubjectivity" (Trevarthan 1979) shows the importance of certain embodied practices – practices that are emotional, perceptual, and nonconceptual – for an understanding of another person's intentions long before the child reaches four years of age. These include the ability of infants to track the other person's eyes and to participate in shared attention behaviours. It is also clear that various movements of the head, the mouth, the hands, and more general body movements are perceived as meaningful (as being goal-directed) and are important for a perceptual (non-conceptual) understanding of the intentions and dispositions of other persons as well as for social reinforcement (see review by Allison, Puce, and McCarthy 2000). There is also evidence for affective and temporal coordination between the gestures and expressions of the infant and those of the other persons with whom they interact. Infants "vocalize and gesture in a way that seems 'tuned' [affectively and temporally] to the vocalizations and gestures of the other person" (Gopnik and Meltzoff 1997, 131). Importantly, the perception of emotion in other people's movement is a perception of an embodied comportment, rather than a theory or simulation of an emotional state. Moore et al. (1997) have demonstrated the emotional nature of human movement using point-lights attached to various body joints. Subjects view the abstract but clearly embodied movement in a darkened room and are able to identify the emotion that is being represented. The emotional states of others are thus not mental attributes that we have to infer. Rather, I perceive the emotion in the movement and expression of the other's body.

Indeed, phenomenologists have a great deal to gain by taking seriously recent discoveries in developmental studies and in neuroscientific research. Studies of "primary intersubjectivity" have a great deal to offer in support of or as corrective of phenomenological analyses. The discovery that newborns are capable of a certain kind of imitation has implications that would push many of Merleau-Ponty's conclusions back to earlier stages of development (see Gallagher and Meltzoff 1996). The discovery of mirror neurons in the premotor cortex and Broca's area has important implications for our understanding of intersubjective perception, as well as the development of gesture and language. All of these scientific studies are suggestive for phenomenology, and, in turn, phenomenology can play an essential role in working out the proper interpretations and implications of these studies.

Rather than simply naming more examples of areas in which phenomenology can work together with the cognitive sciences, we want to turn to an issue that many phenomenologists consider to be fundamental for an understanding of consciousness, namely, the problem of time-consciousness. We want to show that with respect to this specific problem, phenomenology might offer correctives to various cognitive analyses, but also that phenomenology might benefit from some of the more sophisticated cognitive approaches.

Part II: A Resetting of Time-Consciousness

In the cognitive sciences today, studies that address temporality and cognition are usually cast in terms of working (short-term) memory and the "binding" problem (that is, the problem of how the distributed processing of information in the brain can give rise to a unified percept or action). These are specialized problems and for the most part are addressed at levels of sub-personal cognitive mechanisms or in terms of neurological processes. In such cases, the temporality problem seems to be located at a level of analysis that is isolated from more general phenomenological concerns. The notion that the temporality of consciousness is an essential feature that ought to be integrated in a large variety of cognitive analyses is often lost. For this reason cognitive analyses are often static. Furthermore, this forgetfulness of the temporal nature of consciousness leads to accounts of cognition that are phenomenologically problematic.

In this part we want to pursue a twofold thesis. First, we want to show that the phenomenology of time-consciousness can resolve certain problems found in static cognitive accounts of experience. Second, we want to suggest that insights developed in the study of cognitive dynamics can contribute to a better understanding of time-consciousness.

A Cognitive Model and Some Phenomenological Problems

We begin by considering a model of cognition developed by Christopher Frith (1992) in his influential analysis of schizophrenia. Following Feinberg (1978), Frith applies an explanation developed in an analysis of the self-monitoring of motor action to the analysis of mental experience. His hypothesis is that positive symptoms of schizophrenia, such as thought-insertion, delusions of control, and auditory hallucination, involve a failure of self-monitoring which does not occur in normal subjects. Frith provides an account of self-monitoring in terms of subpersonal cognitive mechanisms ultimately intended to be cashed out in terms of neurophysiology. Frith explains the phenomenology of thought insertion in the following way.

> Thinking, like all our actions, is normally accompanied by a sense of effort and deliberate choice as we move from one thought to the next. If we found ourselves thinking without any awareness of the sense of effort that reflects central monitoring, we might well experience these thoughts as alien and, thus, being inserted into our minds. (Frith 1992, 81).

The Feinberg-Frith model assumes that thinking is a kind of action, and that, similar to instances of motor action, it is normally accompanied by an effortful intention. According to this account, the intention to think is the element that guarantees my *sense of agency* for the thought. But

Figure 1: The Feinberg-Frith Model

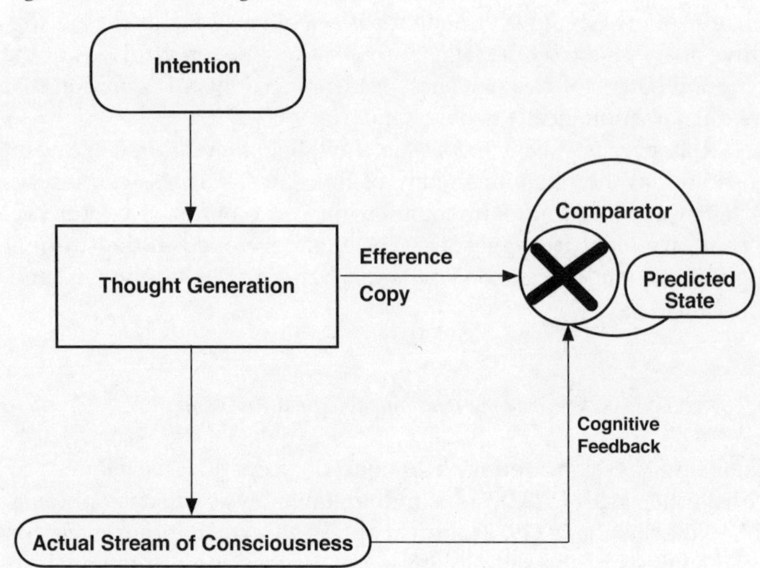

this guarantee works only through mechanisms that operate on a subpersonal, non-conscious level. The intentional generation of thought generates not only the conscious thought, but, on the subpersonal level, an extra signal, an "efference copy," that is sent to a comparator mechanism. The comparator acts as a central monitor, which registers the occurrence of the actual thought, thus verifying that the intention and the thought match (see figure 1).

If the efference copy is somehow blocked from reaching the central monitoring mechanism, thought occurs which seems, to the subject, not to be generated by the subject himself. If the efference copy is blocked or goes astray, or is not properly generated, thinking still occurs, but it is not registered as under the subject's control (intention and thinking fail to match) – it appears to be an alien or inserted thought.

Gallagher (2000) has shown that several aspects of the Feinberg-Frith model are phenomenologically problematic. A first set of problems pertains to Frith's characterization of the intention to think. What role does something like an "intention to think" or the efference copy of that intention play in the case of thinking or conscious experience? It is difficult to conceive of an intention to think prior to thinking itself, unless it is a conscious preparation, as when I might decide to sit down and start thinking about this issue. In that case, however, the intention to think is itself a case of thinking, and we are threatened by an infinite regress: Do I require an intention to think in order to intend to think?

Frith speaks of a *conscious* feeling of effort for a willed intention to think, and he equates this with a *conscious* monitoring of efference copy (1992, 86). This description, which relies not just on an intention to think, but on a "metarepresentation" of the intention to think, fails to capture a built-in sense of agency for thought. Metarepresentation is a second-order reflective consciousness, "the ability to reflect upon how we represent the world and our thoughts." According to Frith, this is part of what it means to monitor our actions and thoughts, and, he claims, it is precisely what is missing or disrupted in the schizophrenic's experience.[5]

Most cases of normal thinking, however, are neither prefaced by conscious intentions to think, nor followed by an introspective metarepresentation. Furthermore, if Frith were to remain consistent with his own model, then metarepresentational introspection is itself a thinking process. It should generate its own efference copy, which would have to be matched on top of the original match. It would thus add an extra level of consciousness to the comparator's verification process and threaten an infinite regress once again.

In the normal phenomenology, at least in the large majority of cases, there is not first an intention and then a thinking, nor thinking plus a concurrent but separate awareness of intention to think. John Campbell (1999a), who considers some version of the Feinberg-Frith account to be the most parsimonious one available, suggests, in contrast to Frith's characterization, that efference copy is not itself available

5 We note that this view of consciousness is somewhat similar to David Rosenthal's "higher-order thought" model (Rosenthal 1997). For a phenomenological critique of this model, see Zahavi and Parnas (1999).

to consciousness. Campbell, however, retains the idea that efference copy matches up with thought itself (introspectively observed) at the comparator in order to verify that the thought is one's own.[6]

It is difficult to understand why something like *efference copy* is really necessary in the case of conscious thinking. In the model of visuo-motor control, from which Frith takes the notion of efference copy, the latter serves a pragmatic or executive function rather than a verificational one. In the case of self-movement, the motor system sends efference copy of a motor command to the visual and vestibular systems, informing those systems to make adjustments, with the very practical effects of stabilizing the visual field. In that case the function of efference copy is to inform the visual and vestibular systems that the organism, rather than the world, is moving. Its purpose is visuo-motor control, not verification that movement is taking place. Is this type of executive function necessary for the thinking process?

Although one can distinguish different cognitive systems – the memory system, the perceptual system, and so on, the Feinberg-Frith model does not assign to efference copy a communicative role among these systems. Campbell suggests, following Feinberg, that efference copy has the pragmatic function of keeping thoughts on track, checking "that the thoughts you actually execute form coherent trains of thought" (1999a, 616). To keep thoughts coherent and on track, however, means keeping them on a semantic track, that is, on a certain track of meaning. Why assign this task to a subpersonal, non-semantic mechanism when,

---

6 Campbell describes the comparator process as involving a form of introspection: "it is the match between the thought detected by introspection, and the content of the efferent copy picked up by the comparator, that is responsible for the sense of ownership of the thought" (1999b). He also states: "You have knowledge of the content of the thought only through introspection. The content of the efferent copy is not itself conscious. But it is match at the monitor between the thought of which you have introspective knowledge and the efferent copy that is responsible for the sense of being the agent of that thought. It is a disturbance in that mechanism that is responsible for the schizophrenic finding that he is introspectively aware of a thought without having the sense of being the agent of that thought."

simply put, we are consciously aware of our thoughts and can keep track of them, and keep them on track, at a conscious level? Nor is it clear that we would ordinarily need a second-order or metarepresentational consciousness to keep our first-order thoughts on track, or to verify that I myself am doing the thinking.

A further set of problems encountered in Frith's model involves its static nature (Gallagher 2000). Frith takes no account of the temporal flow-structure of thought. Although he does acknowledge that the subpersonal comparator mechanisms involve issues of timing (for example, arrival time of efference copy at the comparator relative to registration of the conscious thought), he does not consider the temporal structure of the thinking itself. On a more adequate version of his model, the temporal structure of consciousness would introduce important constraints on the operations of the central monitor.

One motivation for incorporating temporality into Frith's model is to provide an explanation of why the central monitor might fail to register thought as self-generated in some instances but not in all instances. We will refer to this as *the problem of the episodic nature of positive symptoms*. Frith's description of the neurophysiology associated with the positive symptoms of schizophrenia does not address this issue. "Positive symptoms occur because the brain structures responsible for willed actions no longer send corollary discharges to the posterior parts of the brain concerned with perception. This would be caused by disconnections between these brain regions" (1992, 93). One would need to explain why these disconnections manifest themselves in some patients episodically, but not at all times. Simply put, not all of the schizophrenic's thoughts are experienced as inserted thoughts.[7]

---

7 That this is the case is clear, not only from empirical reports by patients, but by logical necessity. The subject's complaint that various thoughts are inserted depends on a necessary contrast between thoughts that seem inserted and those that do not seem inserted – and at a minimum, the thoughts that constitute the subject's complaint cannot seem inserted. If all thoughts were experienced as inserted by others, the subject would not be able to complain "in his own voice," so to speak.

In this case, the phenomenology obviously places constraints on the cognitive explanation.

A second problem for Frith's model involves the selectivity of positive symptoms. In this regard, in cases of thought insertion, specific kinds of thought contents, but not all thought contents, appear to be thought inserted. No explanation that remains totally on the subpersonal level will be able to explain this selectivity. It is not simply that in experiences of thought insertion patients occasionally experience thoughts coming into their minds from an outside source. Rather, their experiences are very specific, and are sometimes associated with specific others. Such thoughts seem to have a certain semantic and experiential consistency that cannot be adequately explained by the disruption of subpersonal processes alone. For example, a schizophrenic will report that thoughts are being inserted by a particular person and that they are always about a specified topic or that, in auditory hallucination, the voice always seems to say the same sort of thing.

A Frithian model that includes the notion of a temporal stream of thought, however, could explain problems with the sense of agency in terms of a lack of temporal synchrony at the subpersonal level – between the stream of thought as represented at the comparator, and efference copy. To explain the episodic nature of the symptoms one could appeal to certain events on the personal or experiential level that would motivate the desynchronization (Gallagher 2000). The personal-level motivation, the description of which is best captured with the help of a phenomenological analysis, might then be understood as part of the explanation for the selectivity of thought insertions.

But even in a less static version of Frith's model, the phenomenological problems involving intention to think and efference copy, as we outlined above, would still remain. In contrast, we want to suggest that a phenomenological model of the sort introduced by Husserl in his analysis of time-consciousness, provides a much more parsimonious account of intentionality and self-monitoring, one, moreover, that would allow an explanation of the episodic and selective nature of inserted thought.

A Phenomenological Model

The Feinberg-Frith model of cognition borrows heavily from explanations of motor control in terms of efference copy and comparators. It is important to note, however, that intentional aspects of motor action, and the generation of the sense of agency for such action, are normally experienced as *intrinsic* to the action. They are phenomenologically indistinguishable properties of the acting itself (Marcel 2003; Gallagher and Marcel 1999). There is good evidence from the study of motor action, for example, that an intrinsic sense of agency is based on "forward," anticipatory processes that occur prior to the action (Georgieff and Jeannerod 1998; Haggard and Eimer 1999; Haggard and Magno 1999). Alain Berthoz, in his recent work on movement, makes much of the ubiquity of such anticipatory mechanisms in the sensory-motor systems. Anticipation is "an essential characteristic of their functioning" and serves our capacity to reorganize our actions in line with events that are yet to happen (Berthoz 2000, 25). The neurological and behavioural evidence suggests that the sense of agency for action, which goes awry in pathological symptoms such as delusions of control, is not based on a *post factum* verification, or a comparator function occurring subsequent to the action or thought. Rather, the sense of agency is generated in an executive or control function that anticipates action.

As we have indicated, efference copy may indeed play an important practical (executive) role in the case of visuo-motor systems, but it is not clear what role it would play in the stream of thought. Alternative and more parsimonious explanations for a sense of agency that is *intrinsic* to thought, and for the loss of the sense of agency in schizophrenic thought insertion, can be advanced by employing Edmund Husserl's model of the retentional-protentional structure of time-consciousness.

My conscious experience includes a sense of what I have just been thinking (perceiving, remembering, etc.) and a sense that this thinking (perceiving, remembering, etc.) will continue in either a determinate or indeterminate way. This phenomenological temporal sense is based on retentional and protentional dynamics, following the logic of time-consciousness that Husserl outlines.

Husserl's analysis of time-consciousness not only explains how the experience of temporal objects is possible, given an *enduring* act of consciousness, it also explain how consciousness unifies *itself* across time.[8] In figure 2, the horizontal line ABCD represents a temporal object such as a melody of several notes. The vertical lines represent abstract momentary phases of an enduring act of consciousness. Each phase is structured by three functions. First, a *primal impression* (pi), which allows for the consciousness of an object (a musical note, for example) that is simultaneous with the current phase of consciousness. Second, a *retention* (r), which retains previous phases of consciousness and their intentional content. And third, a *protention* (p), which anticipates experience which is just about to happen.

Each phase of consciousness involves a retention of the previous phase of consciousness. Since the previous phase includes its own retention of a previous phase, there is a retentional continuum that stretches back through prior experience. There are two important

Figure 2: Time-Consciousness according to Husserl

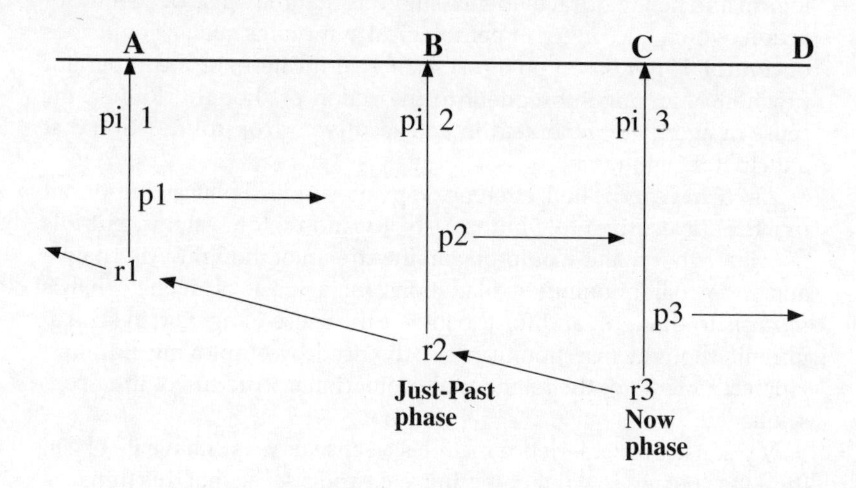

8 A more detailed account can be found in Husserl (1991). For an extended analysis of Husserl's model and its similarities and differences from James's notion of the specious present, see Gallagher (1998).

aspects to this retentional continuity. The first, the "longitudinal intentionality" (*Längsintentionalität*) of retention, provides for the intentional unification of consciousness itself since retention is the retention of previous phases of consciousness. Second, since the prior phases of consciousness contain their respective primal impressions of the experienced object, the continuity of that experienced object is also established. Husserl refers to this as the "transverse intentionality" (*Querintentionalität*) of retention (Husserl 1991, 85).

Retention, for example, keeps the intentional sense of the words of a sentence available even after the words are no longer sounded. Moreover, implicit in this retentional function is the sense that *I* am the one who has just said or heard these words. The words do not become part of a free-floating anonymity, they remain part of the sentence that *I* am in the process of uttering or hearing. Furthermore, if I am in the process of uttering a sentence, I have some anticipatory sense of where the sentence is going, or at the very least, that the sentence is heading to some kind of ending. This sense of knowing where the sentence (the thought) is heading, even if not completely definite, seems essential to the experience I have of speaking in a meaningful way.

The protentional aspect of consciousness provides it with this intentional anticipation of something about to happen. Husserl points out that protention allows for the experience of surprise. If I am listening to a favourite melody and someone hits the wrong note, I am surprised or disappointed. If someone fails to complete a sentence, I experience a sense of incompleteness, precisely because consciousness involves an anticipation of what is to come next, and in these cases, what actually happens fails to match my anticipation. The content of protention, however, is not always completely determinate and may approach the most general sense of "something (without specification) has to happen next."

Husserl's analysis of protention doesn't go much further. As we saw above, however, in his analysis of retention he suggests that there is a double intentionality – one aspect that is directed at the content of experience and another that is directed at consciousness itself. In listening to a melody, I am not only aware of the melody, I am implicitly aware of myself as I am aware of the melody. This implicit, longitudinal intentionality is a non-observational, pre-reflective awareness of my own flowing consciousness, which delivers an implicit sense that this experience is part of my stream of consciousness. This sense of ownership for the experience involves no reflective, second-order, metacognition.

Although protention is asymmetrical with retention in many regards (Gallagher 1998; Varela 1999), there is clearly a longitudinal aspect to protention. That is, my anticipatory sense of the next note of the melody, or of where my sentence is going, or that I will continue to think, etc., is also, implicitly, an anticipatory sense that these experiences will be experiences *for me*, or that *I* will be the one listening, speaking, or thinking. In effect, protention involves a projective sense of what *I* am about to do or experience. Indeed, in contrast to the indeterminate sense of what the content of protention may be, the anticipatory sense of self is relatively determinate.

In the normal case, as we indicated, the *sense of agency* with respect to my own thought does not develop retrospectively, as if I must stop the process and, in a Frithian metarepresentation, think whether I am really the one who is thinking. Rather, taking the clue from the importance of anticipation in motor action, we can pursue the idea that the sense of agency involves an implicit anticipatory moment in thinking itself.[9] As Husserl's analysis shows, this anticipatory aspect is part of the very structure of consciousness, rather than a second-order retrospection or verification. This suggests that protention plays a role in providing a sense of agency in the cognitive domain.

Consider, first, that thought may be generated by the subject in a willed and controlled fashion. Telling a story is a good example. I follow the well-known plot, I have a sense of where I am going in the narrative,

---

9 Following clues from the analysis of motor action, Gallagher (2000) suggests that the dynamics of *protention* underlie the sense of agency for thought, or more precisely, that the protentional registration is a necessary but not a sufficient condition for the sense of agency. The function of *retention*, on the other hand, provides a sense of ownership for thought. A failed protentional mechanism may explain difficulty in generating spontaneous actions, a negative symptom in schizophrenia, as well as difficulties in performing self-directed search (see Frith 1992, 48).

and I push the thinking process along from one step to another in a controlled manner. In this case the protentional aspect of consciousness operates to give me a sense of where my thinking is going *in its very making*, that is, as it is being generated and developed. It provides a sense that the thinking process is being generated in my own stream of consciousness and, to some degree, under my control. Protention normally puts me in the forefront of my thoughts and allows me to take up these thoughts as my own product, as they develop.

A second kind of thinking may be more passive. Unbidden thoughts, memories, fantasies may invade my current stream of consciousness. These are thoughts that I do not intend, although my sense is that these thoughts are coming from myself, rather than from some alien source. They are not only part of my stream of consciousness, but, despite the fact that I am not willing them, and may even be resisting them, they seem to be generated within my own cognitive experience. Within the framework of an unwanted memory or an unwelcome fantasy, protention provides a sense of where the thoughts are coming from and where they are heading, as they are being passively generated. Protention may also provide a sense of *not knowing* where we are heading, a sense of uncertainty, or indeterminacy with respect to where such thoughts will lead. Even in such indeterminacy, I have a sense that they are originating and developing within my stream of consciousness – a sense of *passive generation*.

What would our experience be like if the protentional aspect of consciousness disappeared? In the case of passive, unbidden thoughts, thinking would continue to happen, but there would be neither a sense of agency, nor a sense that these thoughts were being passively generated in my cognitive system, even though they were appearing in my stream of consciousness. Without protention, thought continues, but it appears already made, not generated in my own stream of consciousness. These thoughts would appear as if from nowhere, suddenly and unexpectedly. I would be able to make sense of them only in their retentional train, in retrospect, but not as something self-generated.

This may even be the case with intended thoughts for which I would normally have a sense of agency. Without protention, whatever intention I may have, my sense of what I am about to do or think is disrupted. My non-observational, pre-reflective sense of agency, which I normally experience within a protentional framework, will be missing due to the lack of protention. In this case, I will experience thoughts that seem to anticipate what I *would have* thought. This is consistent with the reports of some schizophrenic patients, that the external force responsible for the inserted thoughts seems to know what they intend to think before they actually think it (Spence 1996). The thought seems to match up with their own intention, but it still seems to them that they are not the agentive cause of the thought.

Without protention, in such cases, thoughts will continue to occur within the stream of consciousness. But they are not experienced *in the making*. Schizophrenics experience what is actually their own thinking, but as thinking that is not generated by them, a thinking that is *already made* or pre-formed for them. This primary failure of protention may then motivate a metarepresentational reflection, an introspection that becomes the hyperreflection characteristic of schizophrenic experience (Sass 1998). In metarepresentation the patient may start to ascribe the seemingly alien thought to some particular force or individual and report that it is inserted (Spence 1996).

Appealing to the Husserlian model of time-consciousness as part of an explanation of intentionality and self-agency for thought, and of their failure in cases of thought insertion, provides a much more parsimonious account than the Feinberg-Frith model. It allows for an *implicit* sense of agency that corresponds to the normal phenomenology of action and thought and the failure of that implicit sense in pathological cases, and it does not require an extra system involving efference copy and comparators. There are still two issues to be resolved. First, can this account be made consistent with the episodic and selective nature of inserted thought. And second, what explains the failure of protention? Both of these questions, however, lead us back to the cognitive sciences.

A Dynamic Model of Protention

There is good empirical evidence to support the idea of a breakdown in the protentional function in schizophrenia. Specifically, research into various aspects concerning schizophrenia and temporality supports this interpretation. These aspects include:

(1) Inability to act toward the future, and a feeling that the future is a repetition of the past (Minkowski 1933)

(2) Difficulties in indexing events in time, with positive correlation to inner-outer confusions (manifested in symptoms such as auditory hallucinations, feelings of being influenced, delusional perceptions, and so forth) (Melges 1982; Melges and Freeman 1977)

(3) Curtailment of future time-perspective (Wallace 1956; Dilling and Rabin 1967)

(4) Difficulty planning and initiating action (Levin 1984)

(5) Problems with temporal organization (Klonoff et al. 1970; DePue et al. 1975)

(6) Difficulties with experienced continuity (Pöppel 1994).

(7) Slowing of temporal processing of different sense modalities, leading to a form of "temporal diplopia" in which consciousness does not seem to coincide with itself. (Pöppel 1994, 192)

These "impairments of self-temporalization" (Bovet and Parnas 1993, 584) are consistent with problems concerning the protentional aspect of experience and are linked with the same neurological dysfunctions involved in the schizophrenic's voluntary movement (Singh et al. 1992; also see Graybiel 1997).

What could cause the failure of protention? There are several ways that one can approach this question. In terms of traditional cognitive science, the task would be to identify a certain set of mechanisms on the subpersonal level. Some aspect of information processing in the neuronal mechanisms that underpin working memory might be responsible for the disruption of protentional consciousness. This approach would be quite consistent with Frith's attempt to identify the cognitive mechanisms responsible for the positive symptoms of schizophrenia. That is, one would attempt to identify specific dysfunctions or disconnections in brain structures responsible for delivery of efference copy to relevant parts of the brain. Such a subpersonal account, however, would necessarily be incomplete unless one could explain why these mechanisms work in some cases but not in others (the problems of the selective and episodic nature of positive symptoms) and how they work with complex psychodynamic processes (such as repression and obsession) in psychopathology.

Insofar as this approach casts the problem in strict information-processing terms (the slowing and discordance of temporal processing) it also ignores an important phenomenological feature of such experience – a feature that may, in turn, play an important role in both positive and negative symptoms of schizophrenia. Clearly, a certain *affective disconcertion* is an integral part of the phenomenological picture associated with these impairments of self-temporalization. That is, the schizophrenic phenomenology is not simply a structural or logical problem that can be captured in terms of a deficiency in information processing. If the subject's experiences can be characterized in some cases as involving a logical "asynchronicity" or a confusion of thought, these cognitive problems are importantly not affectively neutral (even if the content of thought is not emotionally charged) but are characterized by specific affective dispositions, which in some cases involve a lack of or alienation from affect.

This idea links up closely to an explication of protention in terms of *affective tone* (Varela 1999, 2000; Varela and Depraz 2000). On this view, protention is constitutionally involved with an affective tensity or, from a different perspective, a readiness for action. Protention, a formal structure of noetic intentionality, according to Husserl, is directed toward a certain kind of content, which, most formally stated, is *the not yet*. The "not yet" is always suffused with affect at the same time that it is conditioned by the emotional tone that accompanies the flow of experience. Protention involves something like a prediction of the unpredictable. Through protention, one opens oneself to the *not yet* and is in that sense as much self-affected as affected by its content. As Husserl puts it, consciousness "is affected by that which consciousness is conscious of, it trails affect; it is attracted, held, and taken in by that which affects it" (MS C III/I). Depraz explains this in terms of Husserl's emphasis on instinctive intentionality: "Affect is there before being there for me in full consciousness: I am affected before knowing that I am affected. It is in that sense that affect can be said to be primordial" (1994, 75)

We can think of this also in Heideggerian terms: one finds oneself always in a certain disposition (*Befindlichkeit*), lived through as a certain mood, for example, and often this disposition is experientially transparent. That is, one lives through it prenoetically, as a pre-reflective being-in-the-world. At a different level of description, affect has a deeply rooted biological basis; affective disposition is tied to certain neuronal dispositions. Affective tonality, experiential transparency, and readiness for action can be entirely reordered by neuronal events and the balance of neurotransmitters. Such a reordering can sometimes leave a subject specifically indisposed for experience or action.

Is it possible to find a common, albeit abstract, level of description that would capture both the dynamics of neuronal processes and the dynamics of the retentional-protentional flow of time-consciousness? A common feature is that both domains involve self-organizing dynamics (self-constituting processes) which are defined by boundary conditions and initial conditions. Affective tonality can be seen as a major boundary and initial condition for neurodynamics in the temporal flow embodied in brain integration mechanisms (Varela 1999).

A neurodynamical account postulates the following model. Every cognitive act, from perceptuo-motor behaviour to human reasoning, arises through the concurrent participation of several functionally distinct and topographically distributed regions of the brain and their sensori-motor embodiment (Varela et al. 2001). The task of integrating these different neuronal components involves a process that is set out in a complex temporal framework of three different scales of duration (Pöppel 1988; Varela et al. 1991; Varela 1999), the first two of which are directly relevant here.

(1) elementary events (the 1/10 scale varying between 10–100 msecs)
(2) relaxation time for broader integration (the 1 scale, varying from 0.5 to 3 secs)
(3) descriptive-narrative assessments (the 10 scale involving memory)

The various neuronal processes in need of integration at the level of the second scale require a temporal frame or window that defines the duration of the lived present. The best way to understand the recursive structuring of these temporal scales is on the model of nonlinear dynamics.

Evidence for the first scale is found in the so-called fusion interval of various sensory systems: the minimum amount of time needed for two stimuli to be perceived as non-simultaneous, a threshold which varies with each sensory modality. The elementary sensory-motor events that constitute experience (corresponding to what Husserl would call hyletic data) can be grounded in the intrinsic cellular rhythms of neuronal discharges within the range of 10 msec (the rhythms of bursting interneurons) to 100 msec (the duration of an EPSP/IPSP sequence in a cortical pyramidal neuron). These elementary events are then integrated into the second scale, corresponding to the living or "specious" present, the level of a fully constituted, normal cognitive operation. Neuroscience explains this integration in terms of cell assemblies, distributed subsets of neurons with strong reciprocal connections (see Varela 1995; Varela et al. 2001). In terms of a dynamical systems model, the cell assembly must have a relaxation time followed by a bifurcation or phase transition, that is, a time of emergence within which it arises, flourishes, and subsides, only to begin another cycle. Integration occurs because neural activity forms transient aggregates of phase-locked signals coming from multiple regions. Synchrony (via phase-locking) must *per force* occur at a rate sufficiently high so that there is enough time for the integration to hold together within the constraints of transmissions times. In brief, we have neuronal-level constitutive events, which have a duration on the 1/10 scale, forming aggregates that manifest themselves as incompressible but complete cognitive acts on the 1 scale. This completion time is dynamically dependent on a number of dispersed assemblies and not on a fixed integration period; in other words it is the basis of the origin of experienced duration without an external or internal ticking clock. This temporal window is necessarily flexible (0.5 to 3 secs) depending on a number of factors: context, fatigue, sensory modality, age, and so on. This integration-relaxation process at the 1 scale level, corresponds to the living present and allows for an integration that is describable in terms of the retentional-protentional structure.

The kind of self-organization that underlies the emergence of neural assemblies thus involves a component level (at the 1/10 scale) that cashes out in terms of single or groups of nonlinear oscillators. These oscillators enter into a synchrony that is registered as a collective indicator or variable, a relative phase. This collective variable manifests itself at a global level as a cognitive action or behaviour. The self-organization involved here is not an abstract computation, but an embodied behaviour subject to initial conditions (characterized, for example, as what the experiencing subject intends to do or has just done), and non-specific parameters (for example, changes in perceptual conditions, attentional modulation). In this regard there are specific local-global interdependencies. The emerging behaviour or experience cannot be understood independently of the elementary components (for example, hyletic data generated by the organism's interaction with the physical environment); the components attain relevance through their relation with their global correlate.

The fact that an assembly of coupled oscillators attains a transient synchrony and that it happens within a certain temporal window is the explicit substrate of the living present. The dynamical models and the data show that this synchronization is dynamically unstable and will thus constantly and successively give rise to new assemblies (these transformations define the trajectories of the system). Each emergence bifurcates from the previous ones determined by its initial and boundary conditions. Thus the preceding emergence is still present in the succeeding one as the trace of the dynamical trajectory (*retention* on the phenomenological level). The order parameters (initial conditions and boundary conditions) are important here. They are defined by the embodiment and experiential context of the action, behaviour, or cognitive act. The boundary conditions shape the action at the global level and include the contextual setting of the task performed, as well as the independent modulations arising from the contextual setting where the action occurs (i.e., new stimuli or endogenous changes in motivation) (Varela 1999).

The dynamical system described here does not accord with the classical notion of stability that derives from a mechanical picture of the world, or a computational picture of cognition. Stability in the latter case means that initial and boundary conditions lead to trajectories concentrated in a small region of phase space where the system remains, a point attractor or a limit cycle. In contrast, biological systems demonstrate *instability* as the basis of normal functioning – constitutional instabilities are the norm (see Varela 1999 for a summary of the empirical evidence). This instability accounts for the formal flow property of experience. Nonlinear systems provide a self-movement that does not depend (within a range of parameters) on the content of the system. Whether the experiential content of my visual percept is a person or a pyramid, the intrinsic or immanent motion is generically the same, a self-propelled motion. The self-constituting flow of consciousness involves a perpetual change punctuated by transient aggregates underlying momentary acts (at the 1 scale of duration). Changes in initial and boundary conditions drive this flow by motivating transformations to new dynamical phases, in a way that is not predictable along pre-determined trajectories.

Just here, protention plays an important role in the self-movement of the flow. If, as we have suggested, protention is linked to affective tonality (which is reflective of the embodied and contextualized situation), then it helps to define specific boundary and initial conditions for the neurodynamics just described. In the initiation of an intentional cognitive act (for example, I decide to look for a particular object in the environment) I induce a transformation that is coloured by an affective disposition that anticipates the change in perception. In the anticipation of a certain experience, I introduce exogenous order parameters that alter the geometry of the phase space.

Empirical evidence for this can be found in studies of intentional movement. The intention to carry out a movement is coupled with a change in affective tone that varies in degree. One well-known case involves the readiness potential. For a finger movement, a large slow electrical potential can be measured over the entire scalp, preceding by a fraction of a second the beginning of the motion. This is not a correlate of an intention, but it gives some indication of how vast a reconfiguration of a dynamical landscape is involved at the origin of a fully constituted act. Such diffuse effects are in accord with mechanisms associated with neurotransmitters that condition the modes of response at the neuronal level.

On this dynamical view, if protention is linked to affective tonality, then the way to account for the failure of protention is not to search for a particular mechanism (a comparator or misplaced efference copy). Rather, on the neurological level, the sort of mechanism that underlies protention is more appropriately thought of in terms of widely distributed and dynamical processes than in terms of localized functions. As a result, the conceptual framework for thinking about the neurological mechanisms responsible for the symptoms of schizophrenia is quite different from the one involving concepts of comparator, central monitor, efferent copy, etc. Schizophrenic patients feel alienated not just from thought and action; as Louis Sass (private correspondence) points out, they also feel alienated from affects, from their own body and skin, from their own saliva, from their own name, etc. It seems unlikely that all of these phenomena can be explained by problems involving efference copy – problems that may in fact be secondary to a more global dysfunction.[10]

This also means that a disruption of protention is likely to involve widespread cognitive and emotional problems of the sort found in schizophrenia, including incongruity of affect, flat affect (athymia), and "grossly inappropriate affect" (DSM-III-R). Consistent with this picture, premorbid characteristics of schizophrenia patients include difficulties in interpersonal relations, anxiety, neophobia, and defective emotional rapport (Bovet and Parnas 1993).

Importantly, the explanation that links together protentional problems, problems with the implicit sense of agency in cases of inserted thought, and other symptoms of schizophrenia that involve inordinate experiences of time and anomalous affective states, depends on elements in the intentional content of experience, and not simply on disruptions of a subpersonal mechanism. If certain subpersonal

10 Even in the realm of motor action, where such localized mechanisms may be involved in generating the sense of agency, since what is at stake is precisely a *sense* of agency, that is, an *experience* of agency, it is likely, according to the Husserlian model, that the protentional-retentional structure has a role to play in the conscious registration of that sense of agency for movement, or in its failure, for example, in delusions of control.

dynamics are predisposed to malfunction in schizophrenic patients, one possible trigger for this malfunction may be intentional content.

As we saw, a subpersonal explanation does not entirely address the episodic and selective aspects of thought insertion, which may in fact have their proximate cause on the level of semantic/intentional content. There are good arguments and good evidence to show that intentional content has an effect on the temporal structure of experience (Friedman 1990; Gallagher 1998; James 1890). Experience speeds up or slows down according to *what* we are experiencing. Consider, for example, the ordinary experience of how time passes when we are with different people. In some cases time passes too quickly; in other cases too slowly. If boredom can slow the system down and enjoyment and interest speed it up, perhaps anxiety or some experience-related change of affective disposition can cause a disruption of the subpersonal protentional dynamic with a resulting loss of protention in the phenomenological stream. It seems reasonable to propose that a disruption of the protentional dynamic could cause a looping effect that would reinforce the affective trigger. Without protention, for example, it is quite possible that patients would experience others and the world as being invasive, "on top of them," too close, etc., which are, in fact, anxiety-causing experiences, and experiences commonly reported by schizophrenics.

This alternative account of schizophrenic symptoms in terms of a disruption in the protentional dynamic does not need to postulate, in the cognitive domain, comparators or mechanisms involving efference copy. Rather, self-monitoring processes that involve the senses of ownership and self-agency, essential aspects of minimal self-awareness, are built into consciousness as the longitudinal aspects of the retentional-protentional structure. This account requires no mechanisms over and above the mechanisms that constitute the temporal structure of consciousness itself. Furthermore, this account resolves all of the phenomenological problems found in the Feinberg-Frith model. The "intention to think," for example, is not something separate from thinking itself; it is included in the very structure of thought. Accordingly, the schizophrenic does not discover alien thoughts by means of a metarepresentational introspection; rather he will have an immediate, non-observational sense that something is wrong, a sense that is likely to motivate the hyper-reflective metarepresentations that characterize schizophrenia and that lead to further misattributions of agency.

References

Allison, T., Q. Puce, and G. McCarthy (2000). "Social Perception from Visual Cues: Role of the STS Region," *Trends in Cognitive Science* **4**: 267–78.

Baron-Cohen, S. (1995). *Mindblindness: An Essay on Autism and Theory of Mind*. Cambridge, MA: MIT Press.

Bermudez, J., A. J. Marcel, and N. Eilan (1995). *The Body and the Self*. Cambridge: MIT Press.

Berthoz, A. (2000). *The Brain's Sense of Movement*. Cambridge: Harvard University Press.

Bovet, P., and J. Parnas (1993). "Schizophrenic Delusions: A Phenomenological Approach," *Schizophrenia Bulletin* **19**: 579–97.

Campbell, J. (1999a). "Schizophrenia, the Space of Reasons and Thinking as a Motor Process," *The Monist* **82**: 609–25.

Campbell, J. (1999b). "Immunity to Error through Misidentification and the Meaning of a Referring Term," *Philosophical Topics* **26**: 89–104.

Changeux, P., and P. Ricoeur (2000). *What Makes Us Think?* (trans. M. B. DeBevoise). Princeton: Princeton University Press.

Clark, A. (1997). *Being There: Putting Brain, Body, and World Together Again*. Cambridge, MA: MIT Press.

Dennett, D. C. (1991). *Consciousness Explained*. Boston: Little Brown.

Depraz, N. (1994). "Temporalité et affection dan les manuscrits tardifs sur la temporalité (1929–1935) de Husserl," *Alter* **3**: 81–114.

Depraz N. (1995). *Transcendance et incarnation. Le statut de l'intersubjectivité comme altérité à soi chez Husserl*. Paris: Vrin.

Depraz, N., F. J. Varela, and P. Vermersch (2000). "The Gesture of Awareness: An Account of its Structural Dynamics." In M. Velmans, ed., *Investigating Phenomenal Consciousness: New Methodologies and Maps* (121–39). Amsterdam: John Benjamins.

Depraz, N., F. J. Varela, and P. Vermersch (2003). *On Becoming Aware: A Pragmatics of Experiencing*. Amsterdam: John Benjamins.

DePue, R. A., M. D. Dubicki, and T. McCarthy (1975). "Differential Recovery of Intellectual, Associational, and Psychophysiological Functioning in Withdrawal and Active Schizophrenics," *Journal of Abnormal Psychology* **84**: 325–30.

de Vignemont, F. (2000). "When the 'I think' Does Not Accompany My Thoughts." Paper presented at Association for the Scientific Study of Consciousness. Bruxelles (July, 2000). Abstracted in *Consciousness and Cognition* **9**, Part 2.

Dilling, C., and A. Rabin (1967). "Temporal Experience in Depressive States and Schizophrenia." *Journal of Consulting Psychology* **31**: 604–8.

Feinberg, I. (1978). "Efference Copy and Corollary Discharge: Implications for Thinking and its Disorders." *Schizophrenia Bulletin* **4**: 636–40.

Fourneret, P., and M. Jeannerod (1998). "Limited Conscious Monitoring of Motor Performance in Normal Subjects," *Neuropsychologia* **36**: 1133–40.

Franck, N., Farrer, C., Georgieff, N., Marie-Cardine, M., Daléry, J., d'Amato, T., and Jeannerod, M. (2001). "Defective Recognition of One's Own Actions in Patients with Schizophrenia," *American Journal of Psychiatry* **158**: 454–59.

Friedman, W. (1990). *About Time: Inventing the Fourth Dimension*. Cambridge, MA: MIT Press.

Frith, C. D. (1992). *The Cognitive Neuropsychology of Schizophrenia*. Hillsdale, NJ: Lawrence Erlbaum Associates.

Frith, C. D., and D. J. Done (1988). "Towards a Neuropsychology of Schizophrenia," *British Journal of Psychiatry* **153**: 437–43.

Frith, C. D., and U. Frith (1999). "Interacting Minds – A Biological Basis," *Science* **286**: 1692–95.

Gallagher, S. (1998). *The Inordinance of Time*. Evanston: Northwestern University Press.

Gallagher, S. (1997). "Mutual Enlightenment: Recent Phenomenology in Cognitive Science," *Journal of Consciousness Studies* **4(3)**: 195–214.

Gallagher, S. (2000). "Self-Reference and Schizophrenia: A Cognitive Model of Immunity to Error through Misidentification." In D. Zahavi, ed., *Exploring the Self: Philosophical and Psychopathological Perspectives on Self-experience* (203–39). Amsterdam: John Benjamins.

Gallagher, S. (2001). "The Practice of Mind: Theory, Simulation or Primary Interaction." In E. Thompson, ed., *Between Ourselves: Second-Person Issues in the Study of Consciousness*, 83–108. Thorverton, UK: Imprint Academic. Also published in *Journal of Consciousness Studies* **8(5–7)**: 83–108.

Gallagher, S., and J. Cole (1995). "Body Schema and Body Image in a Deafferented Subject," *Journal of Mind and Behavior* **16**: 369–90.

Gallagher, S., and A. J. Marcel (1999). "The Self in Contextualized Action," *Journal of Consciousness Studies* **6(4)**: 4–30.

Gallagher, S., and A. Meltzoff (1996). "The Earliest Sense of Self and Others: Merleau-Ponty and Recent Developmental Studies," *Philosophical Psychology* **9**: 213–36.

Georgieff, N., and M. Jeannerod (1998). "Beyond Consciousness of External Events: A 'Who' System for Consciousness of Action and Self-Consciousness," *Consciousness and Cognition* **7**: 465–77.

Goldman, A. I. (1989). "Interpretation Psychologized," *Mind and Language* **4**: 161–85.
Gopnik, A., and A. N. Meltzoff (1997). *Words, Thoughts, and Theories*. Cambridge, MA: MIT Press.
Gordon, R. M. (1986). "Folk Psychology as Simulation," *Mind and Language* **1**: 158–71.
Gordon, R. M. (1995). "Simulation without Introspection or Inference from Me to You." In M. Davies and T. Stone, eds. *Mental Simulation: Evaluations and Applications*. Oxford; Blackwell Publishers.
Graybiel, A. M. (1997). "The Basal Ganglia and Cognitive Pattern Generators," *Schizophrenia Bulletin* **23**: 459–69.
Haggard, P., and M. Eimer (1999). "On the Relation between Brain Potentials and the Awareness of Voluntary Movements," *Experimental Brain Research* **126**: 128.
Haggard, P., and E. Magno (1999). "Localising Awareness of Action with Transcranial Magnetic Stimulation," *Experimental Brain Research* **127**: 102.
Husserl, E. (1970). *Cartesian Meditations* (trans. D. Cairns). The Hague: Nijhoff.
Husserl, E. (1991). *On the Phenomenology of the Consciousness of Internal Time (1893–1917)* (trans. J. Brough). *Collected Works IV*. Dordrecht: Kluwer.
James, W. (1890). *The Principles of Psychology*. New York: Dover, 1950.
Klonoff, H., C. Fibiger, and G. H. Hutton (1970). "Neuropsychological Pattern in Chronic Schizophrenia," *Journal of Nervous and Mental Disorder* **150**: 291–300.
Leslie, A. (2000). "'Theory of Mind' as a Mechanism of Selective Attention." In M. Gazzaniga, ed., *The New Cognitive Neurosciences*. Cambridge, MA: MIT Press, 1235–47.
Levin, S. (1984). "Frontal Lobe Dysfunction in Schizophrenia – Eye Movement Impairments," *Journal of Psychiatric Research* **18**: 27–55.
Lutz, A., J.-P. Lachaux, J. Martinerie, and F. J. Varela (2002). "Guiding the Study of Brain Dynamics Using First-Person Data: Synchrony Patterns Correlate with Ongoing Conscious States during a Simple Visual Task," *Proceedings of the National Academy of Sciences USA* **99**: 1586–91.
Marbach, E. (1993). *Mental Representation and Consciousness: Towards a Phenomenological Theory of Representation and Reference*. Dordrecht: Kluwer.
Marcel, A. J. (2003). "The Sense of Agency: Awareness and Ownership of Actions and Intentions." In J. Roessler and N. Eilan, eds., *Agency and Self-Awareness: Issues in Philosophy and Psychology*. Oxford: Oxford University Press.
Melges, F. T. (1982). *Time and the Inner Future: A Temporal Approach to Psychiatric Disorders*. New York: Wiley.
Melges, F. T., and A. M. Freeman (1977). "Temporal Disorganization and Inner-Outer Confusion in Acute Mental Illness," *American Journal of Psychiatry* **134**: 874–77.

Merleau-Ponty, M. (1962). *Phenomenology of Perception* (trans. C. Smith). London: Routledge and Kegan Paul.

Merleau-Ponty, M. (1963). *The Structure of Behavior* (trans. A. L. Fisher). Boston: Beacon Press.

Merleau-Ponty, M. (1964). *The Primacy of Perception* (trans. W. Cobb). Evanston: Northwestern University Press.

Minkowski, E. (1933). *Lived Time: Phenomenological and Psychopathological Studies* (trans. N. Metzel). Evanston: Northwestern University Press, 1970.

Moore, D. G., R. P. Hobson, and A. Lee (1997). "Components of Person Perception: An Investigation with Autistic, Non-Autistic Retarded and Typically Developing Children and Adolescents," *British Journal of Developmental Psychology* **15**: 401–23.

Nagel, T. (1974). "What is it like to be a bat?" *Philosophical Review* **82**: 435–50.

Pöppel, E. (1988). *Mindworks: Time and Conscious Experience*. Boston: Harcourt Brace Jovanovich.

Pöppel, E. (1994). "Temporal Mechanisms in Perception." *International Review of Neurobiology* **37**: 185–202.

Port, R., and T. van Gelder, eds. (1995). *Mind as Motion: Explorations in the Dynamics of Cognition*. Cambridge, MA: MIT Press.

Rosenthal, D. (1997). "A Theory of Consciousness." In N. Block et al., eds., *The Nature of Consciousness*. Cambridge, MA: MIT Press.

Rizzolatti, G., Fadiga, L., Matelli, M., Bettinardi, V., Paulesu, E., Perani, D., and Fazio, G. (1996). "Localization of Grasp Representations in Humans by PET: 1. Observation versus Execution," *Experimental Brain Research* **111**: 246–52.

Roy, J-M., Petitot, J., Pachoud, B. and Varela, F.J. (1999). "Beyond the Gap: An Introduction to Naturalizing Phenomenology." In J. Petitot, F.J. Varela, B. Pachoud, and J.-M. Roy, eds. *Naturalizing Phenomenology: Issues in Contemporary Phenomenology and Cognitive Science* (1–83). Stanford: Stanford University Press.

Sass, L. (1998). "Schizophrenia, Self-Consciousness and the Modern Mind," *Journal of Consciousness Studies* **5(5–6)**: 543–65.

Singh, J. R., Knight, R.T., Rosenlicht, N., Kotun, J.M., Beckley, D.J., and Woods, D.L. (1992). "Abnormal Premovement Brain Potentials in Schizophrenia," *Schizophrenia Research* **8**: 31–41.

Spence, S. (1996). "Free Will in the Light of Neuropsychiatry," *Philosophy, Psychiatry, and Psychology* **3**: 75–90.

Thompson, E., ed. (2001). *Between Ourselves: Second-Person Issues in the Study of Consciousness*. Special issue of *Journal of Consciousness Studies* **8(5–7)**.

Tooby, J., and L. Cosmides (1995). "Foreword" to S. Baron-Cohen, *Mindblindness: An Essay on Autism and Theory of Mind* (xi–xviii). Cambridge, MA: MIT Press.

Trevarthen, C. (1979). "Communication and Cooperation in Early Infancy: A Description of Primary Intersubjectivity." In M. Bullowa, ed., *Before Speech*. Cambridge: Cambridge University Press.

Varela, F. J. (1995). "Resonant Cell Assemblies: A New Approach to Cognitive Functioning and Neuronal Synchrony," *Biological Research* **28**: 81–95.

Varela, F. J. (1996). "Neurophenomenology: A Methodological Remedy for the Hard Problem," *Journal of Consciousness Studies* **3(4)**: 330–49.

Varela, F. J. (1999). "The Specious Present: A Neurophenomenology of Time Consciousness." In J. Petitot, F. J. Varela, B. Pachoud, and J.-M. Roy, eds., *Naturalizing Phenomenology: Issues in Contemporary Phenomenology and Cognitive Science* (266–314). Stanford: Stanford University Press.

Varela, F. J., and N. Depraz (2000). "At the Source of Time: Valance and the Constitutional Dynamics of Affect," *Arobase: Journal des lettres et sciences humaines* **4**: 143–66.

Varela, F. J., E. Thompson, and E. Rosch (1991). *The Embodied Mind: Cognitive Science and Human Experience*. Cambridge, MA: MIT Press.

Varela, F. J., J. P. Lachaux, E. Rodriguez, and J. Martinerie (2001). "The Brainweb: Phase-Synchronization and Long-Range Integration," *Nature Reviews Neuroscience* **2**: 229–39.

Wallace, M. (1956). "Future Time Perspectives in Schizophrenia," *Journal of Abnormal and Social Psychology* **52**: 240–45.

Wellman, H. (1993). "Early Understanding of Mind: The Normal Case." In S. Baron-Cohen, H. Tager-Flusberg, C.J. Cohen, and D.J. Cohen, eds. *Understanding Other Minds: Perspectives from Autism* (pp. 10–39). New York: Oxford University Press.

Wimmer, H., and J. Perner (1983). "Beliefs about Beliefs: Representation and Constraining Function of Wrong Beliefs in Young Children's Understanding of Deception," *Cognition* **13**: 103–28.

Zahavi, D. (2001). "Beyond Empathy: Phenomenological Approaches to Intersubjectivity," *Journal of Consciousness Studies* **8(5–7)**: 151–67.

Zahavi, D., and J. Parnas (1999). "Phenomenal Consciousness and Self-Awareness: A Phenomenological Critique of Representational Theory." In S. Gallagher and J. Shear, eds., *Models of the Self* (253–70). Exeter: Imprint Academic.

CANADIAN JOURNAL OF PHILOSOPHY
Supplementary Volume 29

Neurophenomenology and the Spontaneity of Consciousness

ROBERT HANNA AND EVAN THOMPSON[1]

Consciousness is what makes the mind-body problem really intractable (Nagel 1980, p. 150).

My reading of the situation is that our inability to come up with an intelligible conception of the relation between mind and body is a sign of the inadequacy of our present concepts, and that some development is needed (Nagel 1998, p. 338).

Mind itself is a spatiotemporal pattern that molds the metastable dynamic patterns of the brain (Kelso 1995, p. 288).

I. Introduction

It is now conventional wisdom that conscious experience – or in Nagel's canonical characterization, 'what it is like to be' for an organism[2] – is what makes the mind-body problem so intractable. By the same token, our current conceptions of the mind-body relation are inadequate and some conceptual development is urgently needed. Our overall aim in this paper is to make some progress towards that conceptual development. We first examine a currently somewhat neglected, yet fundamental aspect of consciousness. This aspect is the spontaneity of consciousness, by which we mean its inner plasticity and inner

1 We gratefully acknowledge the Center for Consciousness Studies at the University of Arizona, Tucson, which provided a grant for the support of this work. ET is also supported by the Social Sciences and Humanities Research Council of Canada through a Canada Research Chair, and the McDonnell Project in

purposiveness. We then sketch a 'neurophenomenological' framework for thinking about the relationship between the spontaneity of consciousness and dynamic patterns of brain activity as studied in cognitive neuroscience.[3] We conclude by proposing that the conscious mentality of sentient organisms or animals is active and dynamic and that this 'enactive' conception of consciousness can help us to move beyond the classical dichotomy between materialism and dualism.[4]

II. Spontaneity and Subjective Experience

To introduce the spontaneity of consciousness, we first need to delineate the main concepts of consciousness currently discussed in the philosophical and scientific literature. The relationships among these concepts are unclear and the subject of much debate:

1. *Creature consciousness*. Consciousness of an organism taken as a whole insofar as it is awake and sentient (Rosenthal 1997).

Philosophy and the Neurosciences. For helpful comments on portions of earlier drafts, we thank Stephen Biggs, James Campbell, Bryan Hall, Eric Olson, Daniel Stoljar, Jessica Wilson, and the members of the Department of Philosophy at the University of Sheffield. Special thanks are due to the participants in the mini-conference, "Beyond the Hard Problem: Consequences of Neurophenomenology," in Boulder CO, May 2001, particularly David Chalmers and Alva Noë, for intensive discussion of this material and other closely related topics. Finally, we wish to thank the faculty and students at CREA, Ecole Polytechnique, especially Jean Petitot, for stimulating discussion when this material was presented at CREA in March 2003. We dedicate this paper to the late Francisco Varela, who inspired many of its main ideas.

2 See Nagel (1980, p. 160): "An organism has conscious mental states if and only if there is something it is like to *be* that organism – something it is like *for* the organism."

3 On neurophenomenology as a research program in the science of consciousness, see below, section III, and Varela (1996, 1997, 1999); and Lutz and Thompson (2003).

4 The term 'enactive' is taken from Varela, Thompson, and Rosch (1991).

2. *Background consciousness*. Overall states of consciousness, such as being awake, being asleep, dreaming, being under hypnosis, and so on (Hobson 1999). (The coarsest-grained state of background consciousness is sometimes taken to be creature consciousness (Chalmers 2000).)

3. *State consciousness*. Applies to specific and single mental states as individuated by their contents (by contrast with background consciousness) (Rosenthal 1997, Chalmers 2000).

4. *Transitive consciousness* versus *intransitive consciousness*. Object-directed consciousness (consciousness-of) versus non-object directed consciousness (Rosenthal 1997).

5. *Access consciousness*. Mental states whose contents are accessible to thought and verbal report (Block 2001). According to one well-known theory, mental contents are access conscious when they are 'globally available' in the brain as contents of a 'global neuronal workspace' (Dehaene and Naccache 2001).

6. *Phenomenal consciousness*. Mental states that have a subjective-experiential character, that is, there is something 'it is like' for the subject to be in such a state (Nagel 1980; Block 2001).

7. *First-order consciousness*. Unmediated or direct awareness internal to a mental state.

8. *Higher-order consciousness*. The relation between a mental state (not necessarily conscious) and another mental state (such as a thought) that applies directly to the first state.

9. *Introspective consciousness*. Meta-awareness of a conscious state (Jack and Shallice 2001). (Introspective consciousness is often understood as a particular type of access consciousness).

10. *Pre-reflective self-consciousness*. Primitive self-consciousness: Self-referential awareness of subjective experience that does

> not require active reflection and/or introspection (Wider 1998; Zahavi 1999).

Much can be said about each of these notions. A great deal of debate has centred in particular on access consciousness (5) and phenomenal consciousness (6). Some theorists argue that it is possible for there to be phenomenally conscious contents that are inaccessible to the subject (Block 2001); others argue this is incoherent, and hence deny the validity of the access/phenomenal distinction (Dennett 1991, 2001).

This debate gets more complicated when seen from the perspective of phenomenological philosophy (in the tradition of Husserl, Sartre, and Merleau-Ponty). Central to this tradition is the notion of pre-reflective self-consciousness (which goes back to Kant and Descartes).[5] Pre-reflective self-consciousness is a form of primitive self-awareness that is believed to belong inherently to any conscious experience. In other words, conscious experience, in addition to intending (referring to) its intentional object (transitive consciousness/consciousness-of), reflexively manifests to itself (intransitively).[6] Such self-manifesting awareness is a primitive form of self-consciousness, in the sense that it (*i*) does not require any subsequent act of reflection or introspection, but occurs simultaneously with awareness of the intentional object; (*ii*) does not consist in forming a belief or making a judgment;[7] and (*iii*) is 'passive' in the sense of being spontaneous and involuntary. According to phenomenologists, such pre-reflective self-awareness

---

5 For discussion of the Kantian and Cartesian heritage of this notion, see Wider (1998, chap. 1).

6 It is important not to confuse this sense of 'reflexivity' (as intransitive and non-reflective self-awareness) with other usages that equate reflexivity and reflective consciousness (e.g., Block 2001, who regards 'reflexivity' as 'phenomenality' plus 'reflection' or 'introspective access'). As K. Wider (1998) discusses in her clear and succinct historical account, the concept of reflexivity as pre-reflective self-awareness goes back to Descartes and is a central thread running from his thought through Kant, Husserl, Sartre, and Merleau-Ponty.

7 For a recent defence of the idea of a primitive, nonconceptual form of self-consciousness, see Bermudez 1998). For discussion of the connections between Bermudez's account and phenomenology, see Zahavi (2002).

invariably accompanies one's consciousness of objects (outer or inner). A distinction is thus drawn within experience between the 'noetic' or act-aspect of consciousness and the 'noematic' or object-aspect. Experience involves not simply awareness of its object (*noema*), but tacit awareness of itself as act or process (*noesis*). This tacit reflexivity of experience has often been explicated as involving a form of non-objective bodily self-awareness – a reflexive awareness of one's 'lived body' or embodied subjectivity correlative to experience of the object (Merleau-Ponty 1962). Hence from a phenomenological perspective, any convincing theory of consciousness must account for this reflexive experience of embodied subjectivity, in addition to the object-related contents of consciousness (Zahavi 2002).

What bearing does this phenomenological account have on the debate about access consciousness and phenomenal consciousness? According to phenomenology, lived experience always comprises pre-verbal, pre-reflective, and affectively valenced states (processes, events), which, while not immediately available or accessible to thought, introspection, and verbal report, are intransitively 'lived through' subjectively, and thus have an experiential or phenomenal character. Such states, however, are (*i*) necessarily primitively self-aware, otherwise they do not qualify as conscious (in any sense); and (*ii*) because of their being thus self-aware, are access conscious in principle, in that they are the kind of states that can become available to thought, reflective awareness, introspection, and verbal report, especially through disciplined first-person methods of phenomenological attentiveness (Varela and Shear 1999; Depraz, Varela, and Vermeersch 2003).

Whatever might be the ultimate resolution of the debate about access consciousness and phenomenal consciousness, the main point we wish to make here is that none of the above notions of consciousness adequately picks out or characterizes the spontaneity of consciousness (though pre-reflective self-consciousness comes closest). We believe that subjective experience is partially constituted by its being at once underdetermined or uncontrolled by external influences (inner plasticity), and also self-determining or self-controlling (inner purposiveness). It is this dual subjective sense of inner plasticity and inner purposiveness that we mean to indicate with the term 'spontaneity' as applied to conscious experience.

Before presenting a neurophenomenological case for the spontaneity of consciousness, we would like to offer two preliminary phenomenological examples of the spontaneity of subjective experience.[8] The first is the incessant mobilization, demobilization, and remobilization of one's mental resources in the subjective experience of attention. As William James describes it in his *Principles of Psychology*:

> The mind chooses to suit itself, and decides what particular sensation shall be held more real and valid than all the rest.... The mind is at every stage a theatre of simultaneous possibilities. Consciousness consists in the comparison of these with each other, the selection of some, and the suppression of the rest by the reinforcing and inhibiting agency of attention (James 1950, pp. 286, 288).

The second example is the self-generativity of the mind in the subjective experience of imagination. As Edward Casey describes it in his phenomenological study *Imagining*:

> These three characteristics of imaginative spontaneity [i.e., effortlessness, surprise, and instantaneity] may further be regarded as varying expressions of a single basic feature of all spontaneous imagining: its *self-generativity*.... Spontaneous mental acts generate themselves, bring themselves into being.... To be self-generative may mean any of the following: not subject to external coercion or control, but instead bringing *itself* about (thus the "self" of "self-generative" refers not to the imagining subject but to the imaginative act or presentation itself); arising from an apparent lack of cause, motive, or reason; appearing in a way that is unsolicited and unpremeditated by the imaginer and emerging without any express effort on his part; appearing in such a way as to surprise; operating by means of its own self-propelling forces, being wholly self-determinative in this respect; and generating itself all at once, *totum simul*, without any significant prolongation or sense of steady development (Casey 1976, pp. 71–72).

---

8 It is also possible to argue for the spontaneity of consciousness from commonsense functionalism. See Putnam (1975, p. 419).

We think the phenomenology of spontaneity in attention and imagination indicates that the canonical characterization of 'phenomenal consciousness' as 'what-it's-like-for-an-organism' is too narrow. Consciousness is not simply what it's like to *be*, but also what it's like to *do*. To state the point in a more phenomenological idiom, consciousness has an active noetic side in addition to its correlative noematic side: Conscious states need to be individuated not simply in terms of their intentional objects or contents (noemata), but as *intentional acts or processes of experiencing* (noeses). 'Spontaneity' describes the inner plasticity and purposiveness that belongs to the noetic side of experience. In other words, it describes those aspects of consciousness that constitute it as a self-generative or self-organizing (see section IV) process of experiencing.[9]

There are strong hints of this dynamic conception of subjective experience in Kant's and Sartre's notions of spontaneity.[10] The basic idea these two thinkers share is that all experience is, on the one hand, creative or derived in part from (the faculty of) productive imagination, and on the other hand, conative or derived in part from the faculty of desire. Both Kant and Sartre particularly emphasize the extent to which these creative and conative powers of the mind are subsumable under the will or (the faculty of) volition. Hence the upshot of the Kantian-Sartrean notion of spontaneity is that subjective experience is partially constituted by the will.[11]

---

9 Given this active and dynamic character of conscious experience, the term 'phenomenal consciousness' seems unsatisfactory. It carries passive and phenomenalist connotations and is typically used under the problematic assumption that there is a real metaphysical difference between static (fully formed) mental states of subjective phenomenality and subjective accessibility (rather than a notional difference of emphasis and/or a temporal/genetic difference in the emergence of something from pre-reflective to reflective awareness). We therefore prefer the theoretically neutral term 'subjective experience.'

10 See Kant, *Critique of Pure Reason* (Kant 1997), pp. 193 (A51/B75), 205 (B152), and 541 (A548/B576); Sartre (1956, pp. 119–55, 595; and 1987, pp. 79–84).

11 See O'Shaughnessy (2000, chap. 5).

Whereas Sartre focuses mainly on the emotional-affective side of the will, Kant focuses mainly on the rational-intentional side. Spontaneity in our sense, however, is not to be confused with rational, intentional action: Not all subjective experience is self-consciously deliberative or reflective, whereas all rational, intentional action is; and subjective experience in both non-rational, intentional action and unintentional action also exhibits spontaneity. On our view, to say that subjective experience is partially constituted by the will is to say that the sentient organism (as sentient) is intrinsically and always poised to initiate an effective first-order desire or volition, where an effective first-order desire or volition is a desire that moves (will move or would move) the sentient organism all the way to the performance of an action. This sort of desire is to be distinguished from second-order volition, that is, a reflective second-order desire that singles out a certain first-order desire, with the aim of making it effective or volitional.[12] In other words, subjective experience, on our view, is infused with a primitive sense of subject-centred causal power that is logically independent of (though still required by) the richer mental structures of rational cognition, rational agency, and personhood.

III. Neurophenomenology

Before we examine the spontaneity of consciousness from a neurophenomenological perspective, we need to introduce a few of the main ideas of neurophenomenology. Neurophenomenology is a research program for the science of consciousness, proposed by the late Francisco Varela in the mid-1990s.[13] In the experimental context of cognitive neuroscience, neurophenomenology stresses the importance of gathering first-person, phenomenological data from trained subjects as a heuristic strategy for investigating the neural dynamics of

---

12 On the distinction between first- and second-order desires, and between effective first-order desires or volitions and second-order volitions, see Frankfurt (1988, pp. 80–94).

13 See note 3.

consciousness, including its causal efficacy in the brain and organism (Thompson and Varela 2001; Le Van Quyen and Petitmengin 2002). A number of cognitive neuroscientists have recently drawn attention to the need to make systematic use of introspective phenomenological reports in studying the brain basis of consciousness (Jack and Roepstorff 2002). Neurophenomenology takes the further step of incorporating 'first-person methods,' that is, precise and rigorous methods that subjects can use to increase the threshold of their awareness of their own subjective experience from moment to moment, and thereby provide more refined and detailed first-person reports about subjective experience (Lutz, Lachaux, Martinerie, and Varela 2002). The target is to create experimental situations that produce 'reciprocal constraints' between first-person, phenomenological data and third-person, cognitive neuroscientific data: The subject is actively involved in generating and describing specific and stable, phenomenal invariants of subjective experience; and the neuroscientist can be guided by these first-person data in the analysis and interpretation of the large-scale neural dynamics of consciousness. Neurophenomenology thus employs synergistically three fields of knowledge:

1. first-person data from the careful examination of experience with specific first-person methods;
2. formal models and analytical tools from dynamical systems theory;
3. neurophysiological data from measurements of large-scale, integrative processes in the brain.

A pilot experimental study of visual perception using this approach has recently been published (Lutz, Martinerie, Lachaux, and Varela 2002). In this study, new phenomenal invariants in the subjective and cognitive context of perceptual experience were found and described by subjects; and these phenomenal invariants were used to uncover new dynamical patterns of large-scale neural activity correlated with ongoing conscious states.

One of the explicit aims of neurophenomenology, in a conceptual and theoretical context, is to bridge the so-called 'explanatory gap' between the first-person standpoint of subjective experience and the third-person standpoint of cognitive neuroscience (see Roy, Petitot, Pachoud, and Varela 1999). Although the terms 'explanatory gap' and 'hard problem of consciousness' are usually taken to be synonymous, they can be distinguished from the vantage point of neurophenomenology. The explanatory gap, in a neurophenomenological context, is the epistemological and methodological problem of how to relate first-person phenomenological accounts of experience to third-person cognitive-neuroscientific accounts. The hard problem of consciousness, on the other hand, is an abstract metaphysical problem about the place of consciousness in nature. It is standardly formulated as the issue of whether it is possible to derive subjective experience (or 'phenomenal consciousness') from objective physical nature: If it is possible, then physicalistic monism is supposed to gain support; if it is not possible, then property dualism (or substance dualism or idealism) is supposed to gain support (Chalmers 1996). Although Varela (1996) originally proposed neurophenomenology as a "methodological remedy for the hard problem," a careful reading of this paper indicates that neurophenomenology does not aim to address the metaphysical hard problem of consciousness on its own terms. The main reason, following analyses and arguments from phenomenological philosophers, such as Husserl and Merleau-Ponty, is that these terms – in particular the dichotomous Cartesian opposition of the 'mental' (subjectivist consciousness) versus the 'physical' (objectivist nature) – are considered to be part of the problem, not part of the solution. With respect to the explanatory gap, on the other hand, neurophenomenology does not aim to *close* the gap in the sense of ontological reduction, but rather to *bridge* the gap at epistemological and methodological levels by working to establish strong reciprocal constraints between phenomenological accounts of experience and cognitive-scientific accounts of mental processes. At the present time, neurophenomenology does not claim to have constructed such bridges, but only to have proposed a scientific research program for making progress on that task. Whereas neuroscience to date has focused mainly on the third-person, neurobehavioural side of the explanatory gap, leaving the first-person side to psychology and philosophy, neurophenomenology employs specific first-person methods in order to generate original first-person data, which can then be used to guide the study of physiological processes (see Lutz and Thompson 2003).

There are a number of experiential domains in which one could undertake a neurophenomenological investigation of the spontaneity of consciousness. We have already mentioned attention and imagination; to these could be added subjective time consciousness (Varela 1999) and affect/emotion (Varela and Depraz 2000). In the next section we draw from yet another domain – that of the multistable visual perception of reversible or ambiguous figures, such as the Necker cube and Jastrow's duck-rabbit. Our thesis that spontaneity is a fundamental feature of conscious experience is corroborated, we believe, by the phenomenology and cognitive neuroscience of multistable perception. Neurophenomenology in particular enables us to link theoretically the spontaneity of consciousness in the phenomenological domain to the 'metastability' of neural dynamics in the neuroscientific domain. As we shall see, metastability is a dynamical pattern that, at the neural level, seems to mirror the phenomenological pattern of multistability in visual perception.

IV. Spontaneity and Multistable Perception

Psychologists have long been fascinated by visual patterns that spontaneously change or reverse in appearance, such as the well-known Necker cube, named after the Swiss naturalist Louis Albert Necker, who in 1832 reported that line drawings of crystals appeared to reverse spontaneously in depth. The crucial feature of the Necker cube and other reversible or ambiguous figures, such as Jastrow's duck-rabbit, is their perceptual 'multistability' (Attneave 1971): Such figures admit of different 'interpretations' or views; there is a periodic and spontaneous alternation between the different views; and only one view can be seen at any given moment (though there is also a brief interval in which one can experience the reversal or alternation itself). Thus the visual pattern constrains the percept to one of several mutually exclusive alternatives but does not determine which alternative is seen at any given moment, and the percept remains for a time but then spontaneously switches to one of the other alternatives (with a little practice, the switch can also be initiated voluntarily).

There are many different types of patterns that induce multistable visual perception:[14]

(1) Fluctuations in complex geometric patterns, in which innumerable different structures seem to form, break up, and re-form.

(2) Figure-ground tristability: e.g., (*i*) a simple line can be also seen as (*ii*) the boundary of a figure to the right, or (*iii*) the boundary of a figure to the left.

(3) Multistability of symmetry axes: e.g., every equilateral triangle has three-symmetry axes, which determine where the triangle is pointing and where its base is (see Attneave 1971).

(4) Multistability of two-dimensional projections of three-dimensional structures: e.g., the Necker cube.

(5) Multistability of three-dimensional objects: e.g., the perceived direction of movement of most transparent rotating objects switches after some time of observation.

(6) Multistability of motion direction in the apparent motion of two-dimensional visual displays: e.g., stroboscopic alternative motion (left-to-right versus right-to-left) and circular alternative motion (clockwise versus counterclockwise).

(7) Multistability of meaning attribution: e.g., Jastrow's duck-rabbit.

Categories (1)–(6) involve spontaneous reversals in perception. There

14 The following list is taken from Stadler and Kruse (1995). For a different classification of multistable perception, including a wide variety of other examples, see Zimmer (1995).

occurs a transition from one stable perceptual state to another stable perceptual state, with an intermediate phase of instability. This kind of transition, which Stadler and Kruse (1995, p. 6) call 'spontaneous reversion,' has the following basic structure: $Order_1$ -> Instability -> $Order_2$. According to Stadler and Kruse, "Usually the first reversal takes some time, up to three minutes, but then the reversion rate continuously increases until it is stabilized at a rate that is specific to different personalities.... Spontaneous reversions usually do not appear in the multistable patterns of the category (7). An aspect first seen by a person is kept for a long time, maybe forever, if the alternative aspect is not learned" (1995, p. 9).

Four key points about multistable perception are important for our discussion here. First, multistable perception should not be judged as an exceptional or freak phenomenon: "Every stimulus pattern allows more than one interpretation. Thus, every percept has more than one state of stability" (Stadler and Kruse 1995, p. 5). Second, which interpretation of a pattern is seen is accordingly a subjective performance of the perceiver and is not externally controlled by the stimulus. (In phenomenological terms, the noema or intentional object of perception is a function of the noesis or subjective act of perceiving.) Third, in general there are always more than only two aspects available for spontaneous reversion in multistable perception: "Even if, at the first moment, only two aspects are realized, more aspects appear after some time of observation and even further aspects may be learned" (Stadler and Kruse 1995, p. 9; see also Ihde 1986). Nevertheless, as a result of gestalt criteria for perceptual organization, usually not all theoretically possible aspects are seen. Finally, multistable perceptual experiences, because they are dynamic rather than static, have a distinctive temporal character: One experiences an irregular alternation, every few seconds or minutes, between two (or more) different views, including a brief sense of the transition itself, in which the figure appears to jump, as it were, from one aspect to the other. As Varela (1999, p. 270) puts it, "[the] reversal is accompanied by a 'depth' in time, an *incompressible* duration that makes the transition perceptible as a sudden shift from one aspect to the other, and not as a progressive sequence of incremental changes."

Multistable perception provides support for the 'enactive' approach to perception. According to this approach, perceptual content does not arise from the 'recovery' of the external world in an internal representation (after the fashion of orthodox computational vision) or from the direct 'pick up' of information in the ambient light (after the fashion of ecological vision). Rather, it is enacted or brought forth on the basis of self-organizing processes in neural assemblies that, according to the logic of 'circular causality' in complex systems theory, both mediate and are mediated by the sensorimotor loops that embed the perceiver in its environment (see Varela, Thompson, and Rosch 1991, chapters 8–9; Varela 1992; Thompson, Palacios, and Varela 1992). From this perspective, (*i*) external stimuli do not serve to control the perceptual process but act only as boundary conditions on the autonomous (self-organizing) processes of order formation in perception; (*ii*) stability in perception results from fast processes of order formation typically operating on a time-scale beyond conscious awareness; (*iii*) any stimulus condition is potentially multistable; and (*iv*) multistable perception occurs "when the process of order formation (confronted with one boundary condition) spontaneously oscillates between two or more attractors established in the system's dynamics" (Kruse, Strüber, and Stadler 1995, p. 69).

One of the most striking features of the spontaneity of multistable perception is its strong similarity to behavioural and volitional spontaneity. In a recent review article, the neuroscientists D. A. Leopold and N. K. Logothetis summarize the main findings of experimental studies of multistable perception as follows:

> Much evidence suggests that perceptual reversals are themselves more closely related to the expression of a behavior than to passive sensory responses: (1) they are initiated spontaneously, often voluntarily, and are influenced by subjective variables such as attention and mood; (2) the alternation process is greatly facilitated with practice and compromised by lesions in non-visual cortical areas; (3) the alternation process has temporal dynamics similar to those of spontaneously initiated behaviors; (4) functional imaging reveals that brain areas associated with a variety of cognitive behaviors are specifically activated when vision becomes unstable (Leopold and Logothetis 1999, p. 254).

The similarity of the spontaneity of multistable perception to behavioural and volitional spontaneity is also evident phenomenologically.

Careful attention to one's own first-person experience will verify that in perceiving the Necker cube, one's visual presentation will eventually flip from the downwards-left or upwards-right aspect to the other aspect without advance warning or prior conscious intention, and that one cannot prevent this perceptual reversal by any conscious intention, although one can consciously initiate the reversal, especially with practice. Such perceptual spontaneity is strikingly similar phenomenologically to volitional spontaneity or the spontaneity of willing: As one mindfully encounters the world, eventually one will form effective first-order desires or volitions without any advance warning or prior conscious intention, and one cannot prevent this from happening by any conscious intention, although one can consciously initiate desires and volitions, especially with practice. Thus, at the phenomenological level, the spontaneity of conscious experience appears to be constituted by the fourfold fact that the precise, qualitative character of conscious states (1) is not determined by anything external to the conscious subject; (2) is self-generated; (3) is not self-generated by a prior conscious intention; and yet (4) can under some conditions be controlled by a conscious intention.

Of particular importance to us in this paper is the kind of dynamical neural activity involved in the spontaneity of multistable visual perception. Recent neuroscience clearly indicates that visual awareness involves widely distributed brain regions and areas. For an example we can consider recent experimental studies of multistable perception induced by binocular rivalry (see Blake 2001). Binocular rivalry occurs when two different patterns are presented simultaneously, one to each eye. One might think that one would see both patterns, one superimposed on the other. Although one can experience this sort of superimposition at first (depending on various stimulus parameters), one comes rapidly to experience the two patterns as alternating back and forth, competing, as it were, for perceptual dominance. For example, when one eye views a grating moving upwards and the other eye an identical grating moving downwards, one will see a grating that periodically reverses its direction of motion. Binocular rivalry has been used as a tool for trying to dissociate the neural activity that is driven by the stimulus (the pattern before the eye) and the neural activity that corresponds to the percept (the perceptually dominant pattern that the subject reports seeing). In a series of experiments in monkeys, Logothetis, Leopold, and their colleagues found that neural activity at early stages of the visual pathway (primary visual cortex [V1] and visual area 2 [V2]) was better correlated, on the whole, with the stimulus, whereas at later stages the proportion of neurons whose activity correlated with the animal's percept increased, with the highest degree of correlation found in the inferotemporal cortex (IT) (Logothetis and Schall 1989; Leopold and Logothetis 1996; Sheinberg and Logothetis 1997). Nevertheless, "[t]he small number of neurons whose behavior reflects perception are distributed over the entire visual pathway, rather than being part of a single area in the brain" (Logothetis 1999, p. 74). Furthermore, in functional brain imaging studies of human subjects, it has been shown that the transition from one percept to another during rivalry is strongly associated with the covariation of activity in both extrastriate areas of the ventral visual pathway and in frontoparietal areas lying outside the visual cortex (Lumer, Friston, and Rees 1998; Lumer and Rees 1999). For these reasons, "the findings to date already strongly suggest that visual awareness cannot be thought of as the end product of … a hierarchical series of processing stages. Instead it involves the entire visual pathway as well as the frontal parietal areas, which are involved in higher cognitive processing" (Logothetis 1999, p. 74).

Given the distributed character of the neural activity involved in visual awareness – as well as cognitive processes more generally – one needs to ask how this activity is selected and coordinated so as to produce a unified flow of cognitive moments. This issue has recently become known as the 'large-scale integration problem' (Varela, Lachaux, Rodriguez, and Martinerie 2001). One currently explored proposal is that large-scale integration could be mediated by neuronal groups that exhibit a wide range of oscillations (theta to gamma ranges, 6–80 Hertz) and can enter into precise synchrony over a limited period of time (a fraction of a second) (Varela, Lachaux, Rodriguez, and Martinerie 2001; Engel, Fries, and Singer 2001). Synchrony, in this context, means the phase-locking of widely distributed neural populations that behave as coupled nonlinear oscillators. Under conditions of binocular rivalry, for instance, it has been shown (*i*) in cats that neurons whose activity is strongly correlated with the perceived stimulus are strongly synchronized, whereas neurons whose activity is correlated with the suppressed visual pattern exhibit only weak temporal correlation (see Engel, Fries, König, Brecht, and Singer 1999); and (*ii*) in humans, that the coherence (an indirect index of synchrony) between distant brain regions responding to a stimulus is always higher when the subject is aware of the stimulus than when the subject is not (Srinivasan, Russell, Edelman, and Tononi 1999). Direct evidence for long-range synchrony correlated with conscious visual perception in humans has been found by Varela and his colleagues in a study of face recognition (Rodriguez, George, Lachaux, Martinerie, Renault, and Varela 1999). They used high-contrast faces ('Mooney faces'), which are easily recognized as faces when upright, but difficult to recognize when upside down. They found that a consistent pattern of synchrony occurred between occipital, parietal, and frontal brain areas during face recognition of the upright figures, but that this pattern of synchrony was absent when the figures were presented upside down. Moreover, in both cases a new pattern of synchrony in the gamma-frequency range emerged during the motor response given by the subject to indicate perception of the stimuli, and the two synchronous patterns were punctuated in time by a period of loss of synchronization or 'phase scattering.'

To link these points to the spontaneity of multistable perception, we turn to the field of synergetics, a branch of complex dynamical systems theory, which has been especially concerned with multistable phenomena (Kruse and Stadler 1995). In synergetics the brain is treated as a self-organizing system that operates close to instability points in its state space or phase space,[15] and hence is able to switch quickly and flexibly from one state to another. Synergetics models stable perceptual states as attractors in phase space,[16] and multistable perception as a switching between two or more different attractors. Multistable perception occurs near 'saddle node' bifurcations, in which

15 The state space or phase space of a system is a mathematical representation in which each dimension (or axis of the graph) corresponds to one variable in the equation that governs the system's dynamics; thus the space represents all the possible states of the system, i.e., all values of the variables under consideration.

16 An attractor is "an object with no volume in state space toward which all nearby trajectories will converge" (Kellert 1993, p. 10).

two point attractors (valleys) separated by a point repeller (ridge) are simultaneously created when a control parameter reaches a certain critical point.

With these general ideas as background, we can now consider an experimental study by J.A.S. Kelso and colleagues that investigated the dynamics of multistable visual perception (Kelso, Case, Holyrod, Horvath, Raczaszek, Tuller, and Ding 1995; see also Kelso 1995, pp. 218–25). During continuous observation of the same Necker cube figure, a subject's perception will switch back and forth between the two alternative views, and one can record the switching times by asking the subject to push a button every time his or her percept switches. Kelso asked subjects to view a Necker cube in one of eight randomly presented spatial orientations and recorded the switching time for each orientation, thus using the orientation as a control parameter for the switching dynamics.[17] As expected, for each orientation, no consistent pattern was found in the switching-time series data. What Kelso was interested in, however, was whether and how the distribution of switching times might change as the orientation of the cube approached that of a flat figure (which would not be expected to be perceptually ambiguous). He found that the switching time distribution for each orientation was unimodal and asymmetric, with the histograms showing a single hump of varying height and a long tail (a well-known finding for multistable perception), but that as the appearance of the figure approached that of a two-dimensional hexagon (40 deg.) or a square (80 deg.), the frequency distribution was considerably flattened, with the histograms showing an 'extended tail,' indicating that "occasionally a given orientation is perceived for a long time without switching" (Kelso et al. 1995, p. 175). The question he proceeded to address was what kind of underlying dynamics might generate this parameter-dependent distribution.

Kelso's overall theoretical approach combines deterministic and stochastic elements. The non-stochastic element is his dynamical-

---

17 "The orientation of the cube was determined as follows: The cube was assumed to be three-dimensional. The vertical direction was fixed at 30 deg., and the figure was rotated along the polar … direction in increments of 10 deg. ranging from 10 to 80 deg. Zero and 90 deg. rotations result in a 2-dimensional square and so were excluded" (Kelso et al. 1995, pp. 174–75).

systems model of the neural networks involved in multistable perception. In this model, the 'neurons' act as nonlinear oscillators coupled together by a phase relation (the equations that govern this type of system are deterministic). In the language of synergetics, the phase relation is an 'order parameter' or macrolevel collective variable that 'downwardly' constrains the microlevel behaviour of the individual elements of the system (the individual neurons). Order parameters and collective variables are themselves 'upwardly' generated by the microlevel interactions. This two-way (but not symmetrical) determination of microlevel and macrolevel processes is, of course, 'circular causality.'

The stochastic element, on the other hand, comes from the need to recognize the presence of random disturbances in perception:

> As stressed again and again by synergetics, any accurate portrayal of a real world problem must take into account the influence of random disturbances, sources of which in a perception experiment may correspond to factors such as fatigue, attention, boredom, etc. Mathematically, spontaneous switches among attractive states occur as a result of these fluctuations, modelled as random noise. For a given point attractor, the degree of resistance to the influence of random noise is related to its stability, which, in general, depends on the depth and width of the potential well (basin of attraction). As [the control parameter] is increased successively … the stability of the attractor corresponding to the initial percept decreases (the potential well becoming shallower and flatter) leading to an increase in the likelihood of switching to the alternative percept (Kelso et al. 1995, p. 168).

Although spontaneous perceptual reversal is usually assumed to be entirely stochastic (based on random processes), Kelso's dynamical model of the orientation-dependent, Necker cube reversal relies not on an increase of random fluctuations to effect switching, but on the intrinsic character of the model's deterministic dynamics. What the model represents is the temporal evolution of the relative-phase variable (the order parameter), which indicates the global behaviour of the dynamical system. The basic idea of the model is that a fixed point of the dynamical system corresponds to a frequency- and phase-locked state at the neural level, and to a single, unambiguous percept (one view of the Necker cube) at the perceptual level. The distance from a fixed point varies directly with the control parameter (the cube's orientation). As we mentioned, at the nearly symmetric 2-D view (40 deg. or 80 deg.), subjects are more likely to perceive a given orientation for a longer period of time without switching. Similarly, as changes in the coupling parameter of the neurons (the strength of the coupling between them) moves the neural system closer and closer to a mode-locked state, the system resides longer and longer near the fixed point of the relative phase, and the less likely it is to switch.

One of the key aspects of the model – which is central to Kelso's overall theory of neural and perceptual dynamics – is that it exhibits an interesting type of chaotic dynamics called 'intermittency' (technically, so-called Type 1 intermittency), which corresponds to a nearly periodic motion that is interrupted by occasional, irregular, and unpredictable bursts. The bursts appear random, although the system is deterministic. Intermittency provides one well-known route to chaos: As the control parameter increases, the bursts become more and more frequent, until the system is fully chaotic.[18] Intermittency occurs near saddle-node bifurcations (in which a saddle point flanked by two point attractors arises). In Kelso's words: "*The main mechanism of intermittency is the coalescence – near tangency – of stable (attracting) and unstable (repelling) directions in the neural coordination dynamics*" [italics in original] (Kelso 1995, p. 223). In the intermittent regime, the system lives near the boundary, as it were, between regular and irregular behaviour. On the one hand, it tends to hover near the previously stable attractor, as if still attracted to it. Often this phenomenon is described as the attractor's leaving behind a 'remnant' or 'ghost' that still affects the overall dynamical behaviour. On the other hand, the system also occasionally wanders away, as if repelled or ejected, and behaves chaotically until at some later unpredictable time it returns to hover around the ghost of the attractor. With regard to multistable

---

18 "Chaos" is being used here in the technical sense to mean unstable aperiodic behaviour in non-stochastic nonlinear dynamical systems. Chaotic motion is sensitive to initial conditions and impossible to predict in the long run.

perception, Kelso proposes that "the observed switching time behaviour is generated by a coupled, nonlinear dynamics residing in this intermittent régime" (Kelso et al. 1995, p. 182). The idea, roughly put, is that the persistence of a given percept (a given aspect of the Necker cube) corresponds to the system's hovering around the ghost of the attractor; and switching corresponds to the irregular burst, in which the system is ejected or repelled.

We wish to emphasize that this type of unstable behaviour (1) is a generic feature of the complex system; (2) is underdetermined by anything external to the system (e.g., the stimulus configuration and/or control parameter); and (3) is an expression of the spontaneity of the system. This spontaneity corresponds phenomenologically to the plasticity and self-generativity of perception, and neurodynamically to the autonomy or self-organization of the system's dynamics. In Varela's words:

> Complex, nonlinear, and chaotic systems in general provide a self-movement that does not depend (within a range of parameters) on where the systems are. In other words, whether the content of my visual percept is a man or a woman or a pyramid or a hallway [in two different examples of ambiguous figures], the intrinsic or immanent motion is *generically* the same. If the specific place in phase space is a correlate of the intentional content of an object-event, the system never dwells on it, but approaches, touches, and slips away in perpetual self-propelled motion. Cognitively, this corresponds to the observation that in brain and behavior there is never a stopping or dwelling cognitive state [compare this to James's remark about attention quoted in section II], but only permanent change punctuated by transient [neural] aggregates underlying a momentary act.... Formally [in the geometry of phase space], this is expressed by the pervasive presence of stable/unstable regions, so that any slight change in initial and boundary conditions makes the system move to a nearby stable/unstable region (Varela 1999, p. 291).

Given this complex-systems perspective, multistable perception needs to be seen not as a 'molecular' or aggregate experience – an experience compounded out of two or more pre-existing experience-atoms (each view of the Necker cube)[19] – but rather as one 'metastable' experience generated by an integrated and emergent, metastable neurodynamical process (Noë and Thompson 2004). In Kelso's words: "Intermittency means that the perceptual system is intrinsically *metastable,* living at the edge of instability where it can switch spontaneously among perceptual states. Indeed, perceptual states themselves may be metastable, as opposed to corresponding to stable, fixed point attractors. In the intermittency régime, there is *attractiveness* but, strictly speaking, no attractors" (Kelso et al. 1995, p. 182).

This emphasis on instability and metastability as generic features of neural and perceptual dynamics is not idiosyncratic to Kelso's model, or even to synergetics, but is indicative of an important trend in recent neurodynamical research. One of the key notions of this research is that of an 'unstable periodic orbit' (UPO) in phase space. An unstable periodic orbit is a trajectory that visits a large (perhaps infinite) number of periodic orbits, but without ever settling down to one of them: "Instead, the system's behavior wanders incessantly in a sequence of close approaches to these orbits. The more unstable an orbit, the less time the system spends near it. UPOs form the 'skeleton' of nonlinear dynamics, and even the behavior of chaotic systems can be characterized by an infinite set of these orbits" (So, Francis, Netoff, Gluckman, and Schiff 1998, p. 2776). Unstable periodic orbits have been found in different biological systems at many levels, including the human brain (Le Van Quyen, Martinerie, Adam, and Varela 1997). The moral of these findings is that instability or metastability, rather than stability, seems to be the basis of normal functioning in biological systems. In Varela's words: "In this class of dynamical systems, the geometry of phase space needs to be characterized by an infinity of unstable regions, and *the system flows between them spontaneously even in the absence of external driving forces* [our emphasis]. There are no attractor regions in phase space, but rather ongoing sequences of transient visits in a complex pattern of motion, modulated only by external coupling" (Varela 1999, p. 288).

Let us summarize the main points of this section. First, multistable perceptual experiences (e.g., the bistability of the Necker cube) indicate

---

19 We take this molecular analogy from Noë and Thompson (2004).

that perceptual consciousness manifests spontaneity (inner plasticity and inner purposiveness). Second, multistable perception is not an exceptional phenomenon: Every visual pattern allows more than one interpretation, and thus every percept is potentially multistable. Third, which view or interpretation of a pattern is seen is a function of endogenous processes of perceptual-content formation. Fourth, these endogenous processes belong to a self-organizing system whose dynamical activity is metastable. Hence, fifth, conscious experience in perception directly and correspondingly exhibits both neurodynamical metastability and phenomenal spontaneity.

V. Conclusion

Having now accomplished our main aim in this paper – to call attention to the spontaneity of consciousness and to sketch a neurophenomenological framework for thinking about it – we would like to conclude by briefly indicating the broader philosophical perspective on mind and body that informs our discussion, and how this perspective can help us move beyond the procrustean dichotomy of materialism versus dualism in the philosophy of mind.

One important upshot of our neurophenomenological investigation of spontaneity is that conscious experience is enactive – it is a temporal process of active and spontaneous experiencing. More generally, from the perspective of neurophenomenology, the minds of sentient organisms or animals are *enactive* minds. Enactive minds comprise conscious processes (sensory, perceptual, imaginational, emotional-affective, and volitional) that are fully integrated with the self-organizing dynamics of the neurobehavioural processes of living animal bodies, which in turn are both fully embedded in and in constant interaction with their external environments (Clark 1997; Hurley 1998; Juarrero 1998; Thompson and Varela 2001). The spontaneity of consciousness and the neurobehavioural dynamics of the world-oriented animal are thus two sides of the same coin.

Since Hobbes's *De Corpore* and Descartes' *Meditations*, the philosophy of mind has been dominated by the two sharply opposed doctrines of materialism and dualism, which to many have seemed jointly to exhaust the logical space of possible accounts of the mind-body relationship. Roughly, materialism says that the conscious mind is reducible to physical matter, whereas dualism says that the conscious mind is autonomous from physical matter. More technically, 'materialism' can be taken to mean either eliminativism (the denial of the genuine existence of mental properties), or the asymmetric modal dependence of mental properties on physical properties (either as the type-type identity of mental properties with physical properties – whether first-order physical properties,[20] behavioural properties,[21] or functional properties[22] – or as the strong supervenience[23] of mental properties on first-order physical properties). Dualism, on the other hand, can be taken to mean the modal independence of mental properties from physical properties (hence this includes both substance dualism, i.e., Descartes' 'real distinction' between mind and body, and property dualism, i.e., the non-identity of mental properties and physical properties).

The enactive conception of conscious mentality aims to move beyond the classical dichotomy between materialism and dualism in three ways. First, the mind is to be regarded as deeply and inextricably interwoven with an interactive system consisting of brain, body, and world. Second, the mind is emergent from this interactive system in the twofold sense that (*i*) it expresses global properties of the interactive system that do not follow directly from its microphysical components; and (*ii*) it exerts a unique and irreducible causal influence on the local (e.g., microphysical) processes within this system (Thompson and Varela 2001). Third, the intrinsic mental properties and intrinsic first-order physical properties

20 First-order physical properties are properties of the most basic physical entities, processes, and forces.

21 Behavioural properties are mappings from stimulus inputs to operative outputs. These mappings can apply either to machines or organisms.

22 Functional properties are usually construed as second-order physical properties.

23 *B*-properties (i.e., higher-level properties) strongly supervene on *A*-properties (i.e., lower-level properties) if and only if (1) necessarily anything that has some

of animals are at once mutually irreducible and also complementary or necessarily reciprocally related.[24] These three features of the enactive conception, we believe, make room in logical space for a new account of the mind-body relation and a correspondingly new account of intentional action and mental causation. This account is the subject of forthcoming work. We have not presented and argued for this account here but have tried to motivate it through a neurophenomenological account of the spontaneity of consciousness.

---

property *G* among the *B*-properties also has some property *F* among the *A*-properties (or: no two things can share all their *A*-properties in common unless they also share all their *B*-properties in common; or: no two things can differ in any of their *B*-properties without also having a corresponding difference among their *A*-properties), and (2) necessarily anything's having *F* is sufficient for its also having *G*. Strong supervenience can then be fine-tuned by specifying the type of modality or by tightening the relations between higher-level and lower-level properties. Logical strong supervenience, for example, means that the two occurrences of 'necessarily' in the first formulation are to be read as 'logically (or analytically) necessarily,' as opposed, e.g., to 'non-logically (non-analytically, synthetically, strongly metaphysically) necessarily' or 'physically (nomologically, naturally) necessary,' which are more restricted modalities. 'Superdupervenience' means that added 'nomological connections' bind together the *A*-properties and the *B*-properties. See Chalmers (1996, chap. 2); Horgan (1993); and Kim (1993, part 1). It bears repeating that the asymmetric modal dependence involved in materialist identity and materialist supervenience always implies non-reciprocity, i.e., the one-way or 'exclusively upwards' dependence of the mental on the physical.

24 This reciprocal relation is not strictly symmetrical, however, since the structure of the 'bottom-up' relation from first-order physical properties to mental properties is irreducibly and internally distinct from the 'top-down' relation from mental properties to first-order physical properties, just as the structure of the 'local-to-global' or 'upwards' interlevel causal relation from the microphysical parts of neurobiological systems to their emergent macrophysical wholes is irreducibly and internally distinct from the 'global-to-local' or 'downwards' interlevel causal relation from the emergent macrophysical wholes of neurobiological systems to their microphysical parts.

References

Attneave, F. (1971). "Multistability in Perception," *Scientific American* **225**: 62–71.

Bermudez, J. (1998). *The Paradox of Self-Consciousness*. Cambridge, MA: MIT Press.

Blake, R. (2001). "A Primer on Binocular Rivalry, Including Current Controversies," *Brain and Mind* **2**: 5–38.

Block, N. (2001). "Paradox and Cross Purposes in Recent Work on Consciousness," *Cognition* **79**: 197–219.

Casey, E. (1976). *Imagining: A Phenomenological Study*. Bloomington, IN: Indiana University Press.

Chalmers, D.C. (1996). *The Conscious Mind: In Search of a Fundamental Theory*. New York: Oxford University Press.

Chalmers, D.C. (2000). "What Is a Neural Correlate of Consciousness?", in T. Metzinger (ed.), *Neural Correlates of Consciousness*, pp. 18–39. Cambridge, MA: MIT Press.

Clark, A. (1997). *Being There: Putting Brain, Body, and World Back Together Again*. Cambridge, MA: MIT Press.

Dehaene, S. and Naccache, L. (2001). "Towards a Cognitive Neuroscience of Consciousness: Basic Evidence and a Workspace Framework," *Cognition* **79**: 1–37.

Dennett, D.C. (1991). *Consciousness Explained*. Boston: Little Brown.

Dennett, D.C. (2001). "Are We Explaining Consciousness Yet?" *Cognition* **79**: 221–37.

Depraz, N., Varela, F.J., and Vermeersch, P. (2003). *On Becoming Aware: A Pragmatics of Experiencing*. Philadelphia and Amsterdam: Benjamins Press.

Engel, A.K., Fries, P., König, P., Becht, M., and Singer, W. (1999) "Temporal Binding, Binocular Rivalry, and Consciousness," *Consciousness and Cognition* **8**: 128–51.

Engel, A.K., Fries, P., and Singer, W. (2001) "Dynamic Predictions: Oscillations and Synchrony in Top-Down Processing," *Nature Reviews Neuroscience* **2**: 704–16.

Frankfurt, H. (1988). "Freedom of the Will and the Concept of a Person," in H. Frankfurt, *The Importance of What We Care About*, pp. 80–94. Cambridge: Cambridge University Press.

Hobson, J.A. (1999). *Consciousness*. New York: W.H. Freeman.

Horgan, T. (1993) "From Supervenience to Superdupervenience: Meeting the Demands of a Material World," *Mind* **102**: 555–586.

Hurley, S.J. (1998) *Consciousness in Action*. Cambridge, MA: Harvard University Press.

Ihde, D. (1986). *Experimental Phenomenology*. Albany, NY: State University of New York Press.

Jack, A. and Roepstorff, A. (2002) "Introspection and Cognitive Brain Mapping: From Stimulus-Response to Script-Report," *Trends in Cognitive Sciences* **6**: 333–39.

Jack, A. and Shallice, T. (2001). "Introspective Physicalism as an Approach to the Science of Consciousness," *Cognition* **79**: 161–96.

James, W. (1950). *Principles of Psychology*, 2 vols. New York: Dover Publications.

Juarrero, A. (1999). *Dynamics in Action: Intentional Behavior as a Complex System*. Cambridge, MA: MIT Press.

Kant, I. (1997). *Critique of Pure Reason*. Trans. P. Guyer and A. Wood. Cambridge: Cambridge University Press.

Kellert, S. (1993). *In the Wake of Chaos*. Chicago, IL: University of Chicago Press.

Kelso, J.A.S. (1995). *Dynamic Patterns: The Self-Organization of Brain and Behavior*. Cambridge, MA: MIT Press.

Kelso, J.A.S., Case, P., Holyrod, T., Horvath, E., Raczaszek, J., Tuller, B., and Ding, M. (1995) "Multistability and Metastability in Perceptual and Brain Dynamics," in P. Kruse and M. Stadler (eds.), *Ambiguity in Mind and Nature: Multistable Cognitive Phenomena. Springer Series in Synergetics*, Vol. 64, pp. 159–85. Berlin and Heidelberg: Springer Verlag.

Kim, J. (1993). *Supervenience and Mind*. Cambridge: Cambridge University Press.

Kruse, P. and Stadler, M. (eds.) (1995). *Ambiguity in Mind and Nature: Multistable Cognitive Phenomena. Springer Series in Synergetics*, Vol. 64. Berlin and Heidelberg: Springer Verlag.

Kruse, P., Strüber, D., and Stadler, M. (1995) "The Significance of Perceptual Multistability for Research on Cognitive Self-Organization," in P. Kruse and M. Stadler (eds.), *Ambiguity in Mind and Nature: Multistable Cognitive Phenomena. Springer Series in Synergetics*, Vol. 64, pp. 69–84. Berlin and Heidelberg: Springer Verlag.

Leopold, D.A. and Logothetis, N.K. (1996) "Activity Changes in Early Visual Cortex Reflect Monkeys' Percepts During Binocular Rivalry," *Nature* **379**: 533–49.

Leopold, D.A. and Logothetis, N.K. (1999). "Multistable Phenomena: Changing Views in Perception," *Trends in Cognitive Sciences* **3**: 254–264.

Le Van Quyen, M. and Petitmengin, C. (2002). "Neuronal Dynamics and Conscious Experience: An Example of Reciprocal Causation Before Epileptic Seizures," *Phenomenology and the Cognitive Sciences* **1**: 169–80.

Le Van Quyen, M., Martinerie, J., Adam, C., and Varela, F.J. (1997) "Unstable Periodic Orbits in Human Epileptic Activity," *Physical Review E* **56**: 3401–3411.

Logothetis, N.K. (1999) "Vision: A Window on Consciousness," *Scientific American* **281**: 68–75.

Logothetis, N.K. and Schall, J.D. (1989) "Neuronal Correlates of Subjective Visual Perception," *Science* **245**: 761–63.

Lumer, E.D. and Rees, G. (1999) "Covariation of Activity in Visual and Prefrontal Cortex Associated with Subjective Visual Perception," *Proceedings of the National Academy of Sciences USA* **96**: 1169–1173.

Lumer, E.D., Friston, K.J., and Rees, G. (1998) "Neural Correlates of Perceptual Rivalry in the Human Brain," *Science* **280**: 1930–1933.

Lutz, A. and Thompson, E. (2003). "Neurophenomenology: Integrating Subjective Experience and Brain Dynamics in the Neuroscience of Consciousness," *Journal of Consciousness Studies* **10**: 31–52.

Lutz, A., Lachaux, J.-P., Martinerie, J., and Varela, F.J. (2002). "Guiding the Study of Brain Dynamics Using First-Person Data: Synchrony Patterns Correlate with Ongoing Conscious States During a Simple Visual Task," *Proceedings of the National Academy of Sciences USA* **99**: 1586–91.

Merleau-Ponty, M. 1962. *Phenomenology of Perception*. Trans. Colin Smith. London: Routledge Press.

Nagel, T. (1980). "What Is It Like to Be a Bat?" In N. Block (ed.), *Readings in the Philosophy of Psychology*, vol. 1., pp. 159–68. Cambridge, MA: Harvard University Press.

Nagel, T. (1998). "Conceiving the Impossible and the Mind-Body Problem," *Philosophy* 73: 337–52.

Noë, A. and Thompson, E. (2004) "Are There Neural Correlates of Consciousness?" *Journal of Consciousness Studies* **11**: 3–28.

O'Shaughnessy, B. (2000). *Consciousness and the World*. Oxford: Oxford University Press.

Putnam, H. (1975). "The Mental Life of Some Machines," in H. Putnam, *Mind, Language, and Reality: Philosophical Papers, Volume 2*, pp. 408–28. Cambridge: Cambridge University Press.

Rodriguez, E., George, N., Lachaux, J.-P., Martinerie, J., Renault, B., and Varela, F.J. (1999) "Perception's Shadow: Long-Distance Synchronization of Human Brain Activity," *Nature* **397**: 430–33.

Rosenthal, D. (1997). "A Theory of Consciousness," in N. Block, O. Flanagan, and G. Güzeldere (eds.), *The Nature of Consciousness*, pp. 729–53. Cambridge, MA: MIT Press.

Roy, J.-M., Petitot, J., Pachoud, B., and Varela, F.J. (1999). "Beyond the Gap: An Introduction to Naturalizing Phenomenology," in J. Petitot, F.J. Varela, B. Pachoud, and J.-M. Roy (eds.), *Naturalizing Phenomenology: Issues in Contemporary Phenomenology and Cognitive Science*, pp. 1–80. Stanford, CA: Stanford University Press.

Sartre, J.-P. (1956) *Being and Nothingness*. Trans. Hazel Barnes. New York: Philosophical Library.

Sartre, J.-P. (1987) *The Transcendence of the Ego*. Trans. F. Williams and R. Kirkpatrick. New York: Farrar, Straus, and Giroux.

Scheinberg, D.L. and Logothetis, N.K. (1997) "The Role of Temporal Cortical Areas in Perceptual Organization," *Proceedings of the National Academy of Sciences USA* **94**: 3408–3413.

So, P., Francis, J.T., Netoff, T.I., Gluckman, B.J., and Schiff, S.J. (1998). "Unstable Periodic Orbits: A New Language for Neuronal Dynamics," *Biophysical Journal* **74**: 2776–2785.

Srinivasan, R., Russell, D.P., Edelman, G., and Tononi, G. (1999) "Increased Synchronization of Magnetic Responses During Conscious Perception," *Journal of Neuroscience* **19**: 5435–5448.

Stadler, M. and Kruse, P. (1995). "The Function of Meaning in Cognitive Order Formation," in P. Kruse and M. Stadler (eds.), *Ambiguity in Mind and Nature: Multistable Cognitive Phenomena. Springer Series in Synergetics*, Vol. 64, pp. 5–21. Berlin and Heidelberg: Springer Verlag.

Thompson, E. and Varela, F.J. (2001). "Radical Embodiment: Neural Dynamics and Consciousness," *Trends in Cognitive Sciences* **5**: 418–25.

Thompson, E., Palacios, A., and Varela, F.J. (1992). "Ways of Coloring: Comparative Color Vision as a Case Study for Cognitive Science," *Behavioral and Brain Sciences* **15**: 1–74.

Varela, F.J. (1992) "Whence Perceptual Meaning? A Cartography of Current Ideas," in F.J. Varela and J.-P. Dupuy (eds.), *Understanding Origins: Contemporary Views on the Origin of Life, Mind, and Society. Boston Studies in the Philosophy of Science*, vol. 130, pp. 235–63. Dordrecht: Kluwer.

Varela, F.J. (1996). "Neurophenomenology: A Methodological Remedy for the Hard Problem," *Journal of Consciousness Studies* **3**: 330–49.

Varela, F.J. (1997). "The Naturalization of Phenomenology as the Transcendence of Nature: Searching for Generative Mutual Constraints," *Alter* **5**: 355–85.

Varela, F.J. (1999). "The Specious Present: A Neurophenomenology of Time Consciousness," in J. Petitot, F.J. Varela, B. Pachoud, and J.-M. Roy (eds.), *Naturalizing Phenomenology: Issues in Contemporary Phenomenology and Cognitive Science*, pp. 266–314. Stanford, CA: Stanford University Press.

Varela, F.J. and Depraz, N. (2000). "At the Source of Time: Valence and the Constitutional Dynamics of Affect," *Arobase. Journal de lettre et de sciences humain* **4**: http://www.arobase.to

Varela, F.J. and Shear, J. (1999). *The View from Within*. Thorverton, UK: Imprint Academic.

Varela, F.J., Thompson, E., and Rosch, E. (1991). *The Embodied Mind: Cognitive Science and Human Experience*. Cambridge, MA: MIT Press.

Varela, F.J., Lachaux, J.-P., Rodriguez, E., and Martinerie, J. (2001). "The Brainweb: Phase Synchronization and Large-Scale Integration." *Nature Reviews Neuroscience* **2**: 229–39.

Wider, K. (1998). *The Bodily Nature of Consciousness. Sartre and Contemporary Philosophy of Mind*. Ithaca, NY: Cornell University Press.

Zahavi, D. (1999). *Self-Awareness and Alterity: A Phenomenological Investigation*. Evanston, IL: Northwestern University Press.

Zahavi, D. (2002). "First-Person Thoughts and Embodied Self-Awareness: Some Reflections on the Relation between Recent Analytic Philosophy and Phenomenology," *Phenomenology and the Cognitive Sciences* **1**: 7–26.

CANADIAN JOURNAL OF PHILOSOPHY
Supplementary Volume 29

Empathy and Openness: Practices of Intersubjectivity at the Core of the Science of Consciousness

NATALIE DEPRAZ AND DIEGO COSMELLI

1. Situating the Epistemological Background: Beneath the Explanatory Gap

The general framework of this paper relies on the observation that the practice of science as an experimental research program involves a social network of subjects working together (Latour 1988; Stengers 1994), both as co-researchers and as co-subjects of experiments. We want to take this basic observation seriously in order to explore how the objectivity of scientific results obtained thereby is highly affected and dependent on multifarious 'intersubjective regulations.' By intersubjective regulations we mean the different ways in which each subject/researcher is able to account for his or her experience and share it with other subjects/researchers (comparing it, differentiating between each of them, seeing similarities and even identities, but also producing conflictive accounts) to the point of giving way to a re-styled objectivity founded on such ruled inter-individual practices (Husserl 1954, 1973; Depraz 1995; Varela 1999):

> More specifically, 'third-person' protocols are not neutral, that is, true independently of the very situatedness of each subject in its own individuated space and time (Bitbol 2002), but must take into

> consideration 'first-person' accounts and furthermore are inherently dependent on specific 'second-person' validations.[1]

There is today growing evidence in the contemporary cognitive sciences that a disciplined method of collecting first-person data is necessary for the scientific study of consciousness (Varela and Shear 1999). Beyond any sheer phenomenal isomorphism, which only formally correlates subjective experiential accounts and their neuronal counterparts, the neurophenomenological research program as it was initiated by Francisco Varela (1996, 1997) argues that both analyses (neurodynamic and phenomenological) prove to be generated through one another, that is, mutually give rise to new neural data and aspects of the subjective phenomena, on the one hand, and also reciprocally produce renewed dynamical and experiential categories, on the other hand (Varela 1999; Varela and Depraz 2000, 2003).

Relying on this co-generative step, thanks to which first-person singular experiences and subjective accounts, and third-person experimental data and dynamic categories, are mutually constrained, refined, altered, and developed into a renewed, broadened, and deepened unitary field of description, we want to go one step further:

> We propose to take seriously the multifarious varieties of second-person involvements within the whole process of validation.[2] In the long term, we aim at understanding how scientific objectivity is in the last instance radically founded on and determined by such intersubjective practices.

---

1 We explicitly use these denominations 'first'/'second'/'third' person with a still vague meaning, the goal of this paper being to suggest a more precise and renewed understanding of them.

2 By 'validation' we do not understand the mere classic scientific meaning of verifying or controlling the truth of a phenomenon through its observation, but (in a more phenomenologically inspired way) the very 'constitution' of such a phenomenon by an observer who is not neutral or exterior to the phenomenon but completely part of it.

A first step was already taken in that direction in a discussion of the phenomenological pragmatics (Depraz et al. 2003). It was there insisted that the three positions of first person, second person, and third person are a plastic continuum, rather than opposed rigid poles. This discussion thus paved the way for a consideration of the dynamics of the multilayered intersubjective validation of our experience:

> In our view, the second person is not a formal and rigid entity but a relational dynamics of different mediating figures.

Thus we want to radically question the methodological framework of the so-called 'hard problem' as it was first formulated by Nagel (1974) and Levine (1983) and then given its current form by Chalmers (1995). Starting from the irreducible distinction between third-person (experimental and quantified) and first-person (experiential and qualitative) methodologies, and then endeavouring at all costs to bridge the gap leads nowhere: it is from the very start a deadlock. Contrary to such opposed polarities (at least as first- and third-persons are still currently presented), the 'second person' is but a plastic spectrum of interactions: it may be even problematic to carry on calling it a 'second person,' as if we had to do with a separate and isolated entity, whereas it rather corresponds to a relational dynamics in which we are unavoidably immersed. Since the second-person always already appears as such a dynamics of alteration, it will lead us in turn to reconsider the rigid identity of first- and third-persons themselves.

In short, we wish here to promote a renewed 'second-person' methodology with precisely defined criteria and a disciplined procedure as a genuine phenomenological alternative to the so-called hard problem. In that respect, the claim of 'intersubjective regulations' *via* the emphasis put on a social network of subjects-researchers, though it already corresponds concretely to the working and operative scientific framework in a laboratory, needs to be refined further and made into a more precise and explicit methodology.

> Furthermore we put forward the hypothesis that the key Husserlian concept in the analysis of intersubjectivity, namely, empathy, is one

of the two central leading-threads of this disciplined second-person methodology.[3]

Truly we agree with the general claim that empathy is a central condition of possibility of the science of consciousness (Thompson 2001). We would like to go still one step further, however, by appealing to a specific experimental framework in which empathy is put into concrete practice. As a consequence, we will be led to situate empathy at the very *core* of science *as such*. This latter contention is still more radical in two respects: (1) not only the science of consciousness but science as such is affected by empathy as soon as empathy is taken as its leading methodology; and (2) empathy is not only a formal prerequisite of the scientific work (in the sense of a condition of possibility), but it is situated (that is, truly spatio-temporally embodied) in its very middle ground, and this all the more so as it is understood as a concrete practice. In short, intersubjectivity as empathy is not the static foundation of science but the dynamic keystone which holds together its different operations as a *praxis* and generatively regulates its working process. According to such a generative meaning of empathy, which is particularly relevant for the cognitive sciences, we truly need to refine and adjust the Husserlian philosophical concept of *Einfühlung* to the experimental framework, without simply taking that concept for granted. Conversely, the scientific experimental framework needs to be revisited and adapted in the light of such an intrinsic, phenomenological-intersubjective methodology.

We claim that our phenomenological domain cannot be fully understood only from a detached 'the subject behaves like this or that' perspective, or a solipsistic 'I experience this or that' perspective. The nature of such a domain defines its working ground: '*we* experience

3 Insofar as we are looking for more structural and disciplined accounts of intersubjectivity as a methodology, we explicitly part company with other conceptions of Husserlian intersubjectivity, which insist on broader and (to us) vaguer meanings of intersubjectivity, such as Schütz (1967) promotes in relation to the social world, or Dan Zahavi (2001) presents as the intersubjectivity of the world-horizon. We will return to the other meaning of intersubjectivity as 'openness' in Part 3 and how to provide it with a methodological disciplined emphasis.

and we behave,' and calls for a description from within. In doing so it compels us to transcend the subjective-objective dichotomy onto an intersubjective ground that presents itself as more encompassing and fruitful if what we are seeking to describe is the nature of our shared experience: as co-living beings we are bodily coupled, *from the very beginning*, by recurrent interactions (physical, historical and developmental, communicational) and our phenomenology is deployed therein. We bring forth the relevance of empathy as a practice open for development in *this* context. It is indeed a methodological question.

Still one remark before embarking on our study: contrary to some other attempts (Braddock 2001; Gallagher and Cole 2002), we view experience as an upsurging of aspects and dimensions that we could not see at first. In short, our experience is intrinsically generative in the sense that it unfolds by itself new aspects that were first unknown to us. Such an observation leads us to consider how necessary it is for subjects/researchers to be trained in order to be able to grasp in a more satisfying way the surprising complexity of our experience. It also leads us to consider how openness is a complementary key point: it is a fundamental attitude of this disciplined methodology. Our challenge is also that the subjects are truly able to develop a capacity for looking closer at their own experience.[4] It means that experience is not just a given but contains multifarious levels that can be unfolded. In other terms, training unveils aspects of the structure of consciousness and features of my subjective experience that might seem inaccessible and thus might have remained unknown without such a closer examination. Even if the subjects do not need to be full-blown phenomenologists, our contention is that they need to be more than just tuned into their experiences, so as to be able to practice a phenomenological *epoche* of their pre-conceived beliefs and to describe their experience in a disciplined categorized way (Lutz 2002). In the same spirit, science and philosophy are practices that need to be exercised and cultivated. The phenomenological method itself can be refined, made more concrete and put into action (Depraz, 1999b). The convergent claim of all these levels

4 For a first step in that direction, see Lutz et al. (2002). See also as a precursor Stumpf (1883), and in a broader intuitive context, Petitmengin (2001).

of cultivated practices provides the general pragmatic underpinning of our enquiry (Depraz et al. 2003).[5]

2. Intersubjectivity as Experimental Second-Person Methodology

2.1 *A First Step towards a Second-Person Methodology*

In *On Becoming Aware* (Depraz et al. 2003), a first model of the dynamic continuity of intersubjective validation was presented, which goes from the first-person position to the third-person one through two different second-person positions. Putting to the fore a second-person methodology already presupposes that we abandon the notion of closed-up and self-excluding positions and formulate stances constitutively open to other stances.

Thus the three so-called positions involve multiple gradations, which originarily alter the abstract 'purity' of each stance: the so-called third-person position is only ideally objective and neutral insofar as *a socially distributed mediation is always at work*, whereas in turn first-person accounts (Varela and Shear 1999) are far from being private, inaccessible except to the one who experiences: what we experience – what is closest to individual subjectivity – can *also* be examined, expressed, and opened up to intersubjective validation. Far from being polarly opposed to each other, third- and first-person positions both require the introduction of a less obvious position we call the *second person*. It corresponds to an *exchange between situated* individuals, for example focusing on a specific experiential content developed from a first-person position. This could be instantiated by a tutor or guide, someone who has more training in or exposure to a certain domain, and who tries to help the expression and validation of someone else. In short, the three 'positions'

5 Concerning the meaning of phenomenology that is appropriate here, its necessary anchorage in Husserlian phenomenology, but also its renewal and adjustment to contemporary scientific requirements, see Depraz (1999a, 1999b, 2001b).

are not differentiated by the content they address, but by *the manner in which they are inserted in a social network*. Beyond any strict opposition between public and private, or objective and subjective, we thus favour a continuum of intersubjective positions.

Each of these positions therefore offers a particular mode of validation. The third-person position, instantiated in the standard observer of scientific discourse, is the most extreme way of recreating the apparent dualism between internal experience and external objectivity: it is the basis for scientific reductionism in all its forms. This 'pure' form of observation, however, is directly challenged by the cognitive sciences, since their object directly involves the social actors themselves. Thus the cognitive sciences form a singularity in science: no other science has this self-involving structure that co-implicates observer and observed.[6] Although such practices are squarely within the norm of third-person accounts, they already involve the other's position in a way that studying cells and crystals does not. This is the reason that we single it out here as a gradation within the third-person position that edges toward a second-person position.

Dennett has called this very position by the name of *hetero*phenomenology (1991). It is the position of an anthropologist studying a remote culture and inferring models of their members' mental life. Now, in remaining in the anthropological stance, we do not become a member of the tribe. Contrary to Dennett himself, who recently critically mentioned the endeavour of an empathic second-person science of consciousness (Thompson 2001; Depraz et al. 2003), but maintained that every datum can be caught and explored thanks to his heterophenomenological third-person approach to consciousness (Dennett 2002), we nevertheless situate this stance within the range of second-person positions because, even though external traces are used, as situated individuals, we adopt the intentional stance and undertake the interpretations it involves.

What happens when we give in to the temptation to become part of the tribe? In this gradation within the second-person position, we

6 While physics already puts forth the observer as an integral part of the phenomenon, the singularity of cognitive sciences is the self-observation of the observer, and how he or she does so unavoidably immersed in a social context.

observer-interpreters give up some of our detachment and identify with the understanding and internal coherence of our source. An *empathic* transposal into the experiences of the source is then based on our intimacy and familiarity with this type of experience, and on our ability to resonate with others having this type of experience. Even though we do not just melt into the other person, we want to meet on common ground, as members of the group of people who have all undergone the same type of experiences. Moving from the position of anthropologist to that of coach or midwife, whose trades are based on a sensitivity to the subtle indices of the interlocutor's phrasing, body language, and overall expressiveness, serves as an inroad into a common experiential ground. Empathic resonance is thus a radically different style of validation.

But the second-person stance can also be seen from the point of view of the source, the person undergoing the experience to be examined. Here we focus on you as a subject, who, having decided to seek a validation for an expression, moves into an examination session in which you submit to the mediation of another. At some point, in order for it to be fruitful and socially engaged, all first-person work must eventually, sooner or later, assume the position of a direct experience that refuses continued isolation and seeks intersubjective validation.

Finally, leaving to one side the issue of delusional isolation, the first-person position contains a form of internal validation, which draws its force from the nature of intuitive fulfillment. This experience, while it is certainly not incorrigible, does at least hit you with an *aesthetic* force. The immediacy of fully accomplished intuitive insight, then, is the most 'pure' subjective form in the spectrum of validation, but we must be clear that it has this place only to the extent that it *belongs* to that continuum: it is not in any way isolated from the other types of validation available to the other gradations of first-, second-, and third-person positions.

2.2 *Beyond On Becoming Aware*

On such a basis we want to show how such a first framework, though structurally relevant, still needs to be refined and differentiated, while detailing multifarious empathetic second persons, which appear to be strongly irreducible both to first- and third-person stances, even in the way we were able to revisit them as we did first. Since Dennett's heterophenomenology obviously stands close to our project, we want to clarify further our contention with regard to it. To this end, we give a first, preliminary account of empathy and the second-person perspective, which will help to situate our approach in relation to Dennett's. We then clarify the points of contention between us and Dennett. This clarification will in turn lead us to differentiate once again, but now more precisely, the inherent plasticity contained in first- and third-person stances themselves.

2.2.1 The differentiated plasticity of second-person stances proper: A phenomenological approach

Introduction

In this first step dealing with a phenomenological intersubjective/second-person methodology *in a singular experimental framework*, we would like to present empathy as a central working experience. In that respect we adopt a critical stance against the notion of an intersubjectivity that would go 'beyond empathy' as if the project were to leave out empathy in the direction of a more encompassing understanding of intersubjectivity merely involving the three regions of self, others, and world (Zahavi 2001).

What do we mean by empathy?

It is possible to distinguish four different and complementary stages of empathy:

- (*i*) a passive association of my lived body with your lived body
- (*ii*) an imaginative self-transposal in your psychic states
- (*iii*) an interpretative understanding of yourself as being an alien to me

(*iv*) an ethical responsibility toward yourself as a person (enjoying and suffering)

i. Coupling (*Paarung*)

Empathy is not first and foremost conditioned by my visual perception of the body of the other, which would mean that we mostly have to do with the meeting of two perceptual and reflecting body images. Empathy is grounded in a much more passive and primal experience lying in both our lived bodies (in our body schemas). Husserl has a name for such a hyletic underground: he speaks of *Paarung* (coupling). Coupling is an associative process through which my lived body and your lived body experience a similar functioning of our tactile, auditory, visual, proprioceptive body-style, of our embodied behaviour in the world, and of our affective and active kinesthetic habits and acts. Coupling is a holistic experience of lived bodily resemblance.[7] It is the grounding process of empathy, without which no further intersubjective experience is possible, be it the experience of dissimilarity (pathological or not), or of focusing on one aspect of the body (face-to-face or shaking-hands experiences).

ii. Imaginative Self-Transposal (*sich Hineinphantasieren*)

Having experienced such a global resemblance of our body-style, I quite spontaneously transpose myself imaginatively in yourself. You are assaulted by unpleasant rememberings, pleased with some daydreaming, worried about some dark feelings. You tell me about such psychic states and I immediately transpose them as being possibly mine. I recall similar experiences where I had such mental states and I am then able to feel empathy. 'Imaginative self-transposal' deals with the 'cooperative encounter' of our embodied psychic states, as Spiegelberg named this second stage of empathy after Husserl (Spiegelberg 1971, 4; 1995). Again, Husserl has a name for such a second stage. He calls it *sich Hineinphantasieren*. I am here and I imagine I am going there

7 We stress the fundamentally historical nature of such coupling.

to the place where you are just now; conversely, you are here (the 'there' where I am going) and you imagine you are going there, to the place where I am (my 'here'). Literally, we are exchanging places at the same time: through imagined kinesthetic bodily exchanging we are able to exchange our psychic states. Such a second stage is highly embodied because it relies upon a concretely dynamical spatializing of imagining.

The two further steps, (*iii*) understanding and communication, and (*iv*) ethical responsibility, can be summarized as follows: the third step of empathy involves expression (verbal or not) and interpretation, which lead to the possibility of understanding (and misunderstandings of course): it is a cognitive step. The fourth step deals with ethics and affection and considers the other as having emotions – as suffering, enjoying. In that respect, Scheler (1970), with his multifarious descriptions of sympathy in *The Nature of Sympathy* is far more helpful than Husserl.

These four stages are not chronological but correspond to what Gendlin called a 'logic of experiencing' (Gendlin 1997). They occur together in our intersubjective experience, which is at each step structured by what Husserl calls a lived analogizing (*Analogisierung*): at the very moment when the other experiences and constitutes me as a body similar to his/her own body, I experience and constitute him/her as a body in this very same way similar to mine (external similarity). Yet, while constituting me as a body, the other discloses to me my body as an object, which I inhabited as a habitual and unreflexive lived body; and I, by constituting the body of the other as a habitual and unreflexive lived body, disclose to the other his/her body as a subject (inner similarity).

Intersubjectivity thus appears to be a *mutual discovery*:[8] through the other's own embodiment, I become fully embodied and aware of the constitutive effacement and forgetfulness of my own functioning lived-body, while the other simultaneously acquires such a self-awareness.

8 And in this sense fundamentally co-generative.

Examples

Let us consider some first-person examples (the 'I' in following two examples refers to Depraz):

> I start reading Antoine's latest version of his article about depth perception. First I just go through the content, as it is quite familiar to me, as if reading through the lines. Then I suddenly discover a sentence which gives a genuine sum-up of his position. To and fro I am making notes about sentences which would need clarification to me, some others of which I am not sure I grasp their full meaning. I gradually begin to discuss with the text, asking questions in the margin, agreeing with some points, having reservations about some others. The text is now literally 'talking' to and with me.

Such a dialogue with a text is in fact a discussion between an I and a You, in an interpersonal mode that is both emotional and cognitive: there is an attraction towards the ideas of the other, which creates the interest and wish to read further. In this respect, empathy is inextricably affective and theoretical, in a mixture that is difficult to disentangle.

> Wednesday afternoon in my 'phenomenological workshop.' We meet the five of us with the scheduled aim of beginning to sort out our different descriptions of Diego's experiment on 'binocular rivalry.' First Diego starts with a scheme he presents to us, showing the different relationships between him and the 'subjects,' himself being a subject, we in turn taking part actively in the experiment as co-researchers. He also mentions the different descriptive trials for each of us, the registered ones (on tape or as encephalogram). A question arises (Juan) around the meaning of what is a 'phenomenological reduction' in these descriptions: to what extent is it at work in them? How? A detailed discussion comes about as far as the relevance of the understanding of reduction is concerned, and how to make a clear distinction between the contents of the description and the way the subject describes it. Another question then broaches the issue of the general disposition of the subject (Jean-Philippe) at the beginning of the experiment, of Diego's role in not inducing any explicit search in the subject, that is, in his letting-go attitude (Natalie). We are reminded (Jean-Philippe) of the three

steps of the epoché as they are laid out in *On Becoming Aware* (Depraz et al. 2003). But the question still is: how and when do you describe? What do we call 'descriptive categories,' 'features' (Claire). Aren't they too abstract? (Diego) The question remains open at this point and we decide to proceed first to a new examination step-by-step of each of our descriptions with Diego, with the task of isolating the content from the way it is described.

As the discussion develops questions arise that were not foreseen at first. So the collective dynamics is an interplay of known facts and unknown dimensions of the experience: it is a genuine path of discovery insofar as none of us knows in advance the whole of the matter. Each step corresponds to the potential opening of a new aspect that could not be seen as such.

A first scheme: Multiple second persons

To summarize the discussion up to this point, we schematize the differentiated plasticity of second-person perspectives in Figure 1.[9]

[Figure 1: multiple second persons]

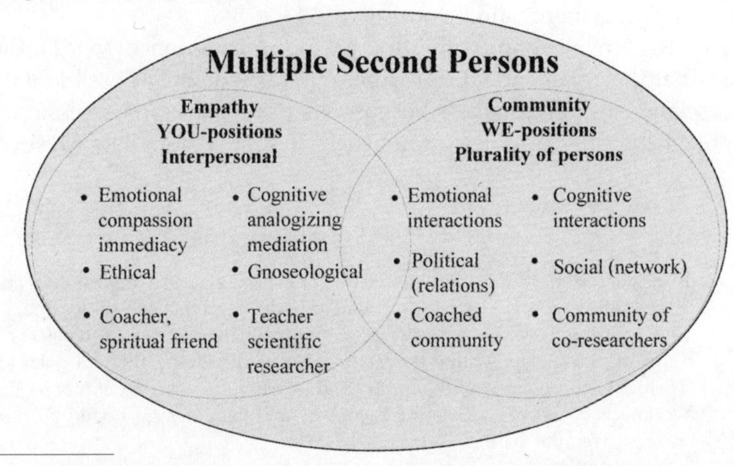

9 The use of schemas tends to overstress an analytical attitude and therefore an abstract separation of categories. We nevertheless point to the heuristic richness

2.3 *Dennett's Contention*[10]

With the heterophenomenological research program (Dennett 1991), we have to do with a method where the distinction between the first-person experience and the beliefs we have about this very experience is central. Dennett's clear contention is that only these beliefs about our experience (and not our experience as such) are reliable and furthermore useful as data in order to study our consciousness in a scientific manner. The goal is to use all the possible technical tools (interviews, recorders, etc.) in order to collect as systematically as possible the whole of the beliefs that a subject has about a given experience, to the point of making of these beliefs a coherent system.

In that respect, we can make a preliminary distinction between two different notions of experience (1) experience is our beliefs about experience, which means that experience is always already filtered by preconceptions (in the spirit of Freud and Derrida, for example), and *interpretations* are our only accessible reality; (2) intuitive first-person experience has a genuine character, which can be accessed by any individual. In order to explore adequately this genuine character of experience requires the development of an attention towards oneself, which implies that experience is a *trained* experiencc (following Husserl, Spiegelberg, and Gendlin).

Dennett (1991) upholds the first notion of experience, that is, the view that the beliefs about our experience are the only possible data of a science of consciousness because we have conscious access only to them: the flow of our consciousness is a flow of beliefs (desires,

---

of such descriptions that allow us to see all poles in relation to one another and we stress this aspect of it. It is important to understand that such poles operate more as attitudes we embody during our daily life than simply fixed objects of study. We must keep in mind the relational dynamics of the different poles we lay down and understand that no fruitful investigation of such dynamics can overlook this aspect by isolating them first and then trying to build abstract bridges to reunite them.

10 We wish to thank O. Sigaud for providing us with relevant insights about this discussion of Dennett's heterophenomenology and our practical phenomenology.

intentions), and it cannot be presupposed, Dennett argues, that the objects of such beliefs really exist (they might also be imaginary). So in a way similar to the Husserlian epoché, Dennett brackets the existence of the objects of the beliefs and is interested only in the intentions themselves. Another way to understand Dennett's position would be to say that the beliefs about our experience are the only scientific data that we can rely on. There might be other forms of access to our experience, but only via beliefs can we access our experience. This is closer to a phenomenological approach because the statement includes that we have a genuine experience and an access to it but makes the point that we need to go through the formation of beliefs (what Husserl would call sedimentation and habitus) in order to gain access to the experience.

But the question then is: if we can make a distinction between a genuine experience and beliefs about it, it is tempting to ask if we can access this genuine experience without going through its beliefs. This question also bears on the genetic process of sedimentation and habitual belief formation out of our primary genuine experience. Husserl's contention lies precisely in such a possibility of a direct intuitive access to our genuine experience ('static constitution'), and in the description of the genetic production of our sedimented beliefs ('genetic constitution'). Husserl's science of conscious experience starts from the postulate that we have an apodictic (non-doubtful) access to our own experience and that we can observe, examine, and describe it accurately. His strong criticism of skepticism (Husserl, 1950a, §76–79, p. 247) is motivated by a main hypothesis: reflection is an operation that provides us with genuine access to our lived experience ('vécu' in French). Or again: every lived experience can be examined through reflection (Husserl 1950a, 248). The question is: does reflection as an operation modify or alter my lived experience?

Some conclusions can be drawn from such a strong stance: (1) there is a differentiation between lived experiences that are inherently intentional and reflective and others that are not, as in the example of merely lived, un-reflected joy (Husserl 1950a, 247; Nam-In-Lee 1998); (2) another distinction lies in the understanding of consciousness as direct consciousness or as reflecting consciousness, inasmuch as intuitive evidence is concerned; (3) a third point has to do with the necessity to rely on training in order to unfold intuitive evidence, which nothing can actually argue for, except the strong inner feeling of evidence, a priori justified in apodicity. In this respect, it seems to be necessary to add the notion of phenomenological training to Husserl's thesis (as Stumpf already did).

2.4 *Implications for First-Person and Third-Person Stances: Extended Differentiations*

The discussion of Dennett's contention brings about further differentiations inside the first-person approach on the one side, and inside the third-person approach on the other side. On the one hand, a subject may consider him or herself in two different ways as a self: (1) as being intimately related to him or herself, with an inner search for transparent, that is, apodictic, truth of what he or she is; (2) as being aware of his or her ceaseless self-interpretative strategies towards him or herself, which unavoidably leads to 'split' conscious states (in a sense to be clarified below). On the other hand, from a third-person point of view, one may look at others as potential selves like oneself or radically observe them neutrally from the outside.

In short, in contrast with the four different 'second-person' polarities already mentioned, we can distinguish between at least two 'first' persons and two 'third' persons, these seven poles forming together a strong continuity of many 'intersubjective' stances in a broad sense of the word. This differentiation of multiple second persons and renewed first persons and third persons is schematized in Figure 2.

[Figure 2: A second scheme: multiple second persons and renewed first persons and third persons]

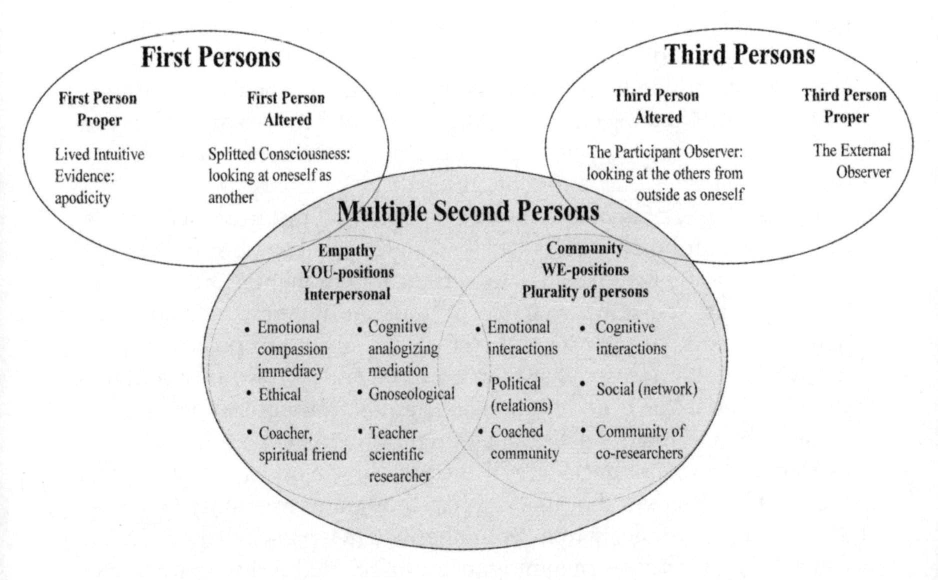

Self-alterity

Let us unfold what is at stake in the second understanding of the first-person pole as a split, self-interpretative consciousness. In order to do so, we need to delineate three main concepts – namely, intentionality, self-alterity, and self-alteration – the first one being strictly Husserlian, the two others belonging to the interpretation of Husserl's phenomenology of intersubjectivity (Depraz 1995).

Husserl's first main discovery amounts to the suggestion of a new understanding of consciousness. The features of consciousness are no longer Cartesian, that is, inner, closed, substantial, and solipsistic. Consciousness, for Husserl, is intentional, which does not mean first and foremost deliberate or wilful, but rather simply a *directedness* towards the external object and an *openness* to the world. In that respect, intentionality can be active or passive, voluntary or driven, attentional or affective, cognitive or emotional, static or genetic. Furthermore, open-directedness indicates a strong relativization of the subject-object polarity. Heidegger's original contribution to this new concept of consciousness will consist in making such a relativization still more radical by deepening intentionality into a *self-transcendence* of *Dasein,* the latter being sheer openness to the *world,* without any residue of a subject. As for Merleau-Ponty, his emphasis on the *lived body* will lead him to a pre-conscious, anonymous and strictly functional *immanent-embodied* intentionality.

On the basis of such a renewed understanding of consciousness through intentionality, self-alterity suggests both a deepening and a change of stress: whereas Husserlian intentionality and *a fortiori* Heideggerian self-transcendence or Merleau-Pontian world-immanence are external openings of our consciousness to objects and to the world, self-alterity consists in an *inner opening* of our egoic subjectivity. The latter is no longer a substantial unity but appears to be ceaselessly and innerly self-altered. In most of my experiences, which as intentional experiences entail meeting an object or encountering the world, I discover that this 'I' I had thought of as a unity has been altered: for example, while remembering, my present I is different from my past I; while imagining, it could be another I than my effective I; while reflecting it is not the same as my reflected I and it is not the felt I similar to my feeling I.

The primal self-alterity of the ego is based on intentional/transcendent/embodied consciousness as its experiential condition but discloses the egoic subject itself as being innerly inhabited by such many egoic splittings. Self-alterity is not a unified concept. That would be contradictory to its very meaning, for self-alterity corresponds to such multifarious experiences of the I that its very significance lies in its non-unifiability. There is, however, no complete bursting of the ego in self-alterity: the ego remains itself self-altered.

Self-alterity provides a fine tool to understand better that there is no exclusive alternative between ego and non-ego. The truth lies in a middle path, which can be called a self-altered ego. In that respect, 'alterology,' being the science of such a self-altered subjectivity, constitutes an inner alternative to egology.

Still, self-alterity remains static. It needs to be complemented by the dynamics that underlies it, namely, self-alteration. After having remembered, imagined, reflected, felt, I may become conscious of these inner alterations of myself. Such a consciousness of having been altered becomes pregnant only in the aftermath. But I may also become aware of self-alterations of myself at the very moment when they are occurring. In that case, I am able to attend their very birth in me, and then, their genesis in me. I may then be able to observe the passive, associative, affective, habitual emergence of my lived embodied consciousness. Husserl calls such a genesis of egoic subjectivity a hyletic (or material) genesis. *Hyle* is a synonym for passive, associative, affective, driven, and habitual and is frequently defined as *das Ichfremde*, namely, what is foreign to the I but deeply inhabits me: in short, the alien within myself and inherent in me.

It is interesting to observe in this connection that living beings, as recurrent processes of self-assertion (Varela 1979, 1991), bear intimately the very definition of what they are not. Self-assertion is not just a one-time act; it has to be renewed: our internal dynamics is that of making oneself, constructing one's identity at each moment, while stopping means dying. At each instant I have to actively sustain the identity that escapes me, I have to grasp that which I do not have in any tangible way in the face of what I am not. The deep vertigo towards *what is not me* has indeed its roots in the very bio-logic we incarnate as living beings.[11]

We thus arrive at a third scheme of further differentiations, in the form of extended first persons and third persons (Figure 3).

11 This approach is strongly inspired by Hans Jonas's *The Phenomenon of Life* (Jonas 1966).

[Figure 3: A third scheme: further differentiations; first persons and third persons extended]

First Persons

First Person Proper
Lived Intuitive Evidence: apodicity

First Person Altered
Splitted Consciousness: looking at oneself as another

Third Persons

Third Person Altered
The Participant Observer: looking at the others from outside as oneself

Third Person Proper
The External Observer

Self -alterities/alterations

1. Myself as a past I/present I
2. Myself as an imagined I/effective I
3. Myself as a reflected I/reflecting I

(Husserl, 1973; Depraz, 1995)

Other -altering processes

1. First towards second person: self-subjectivation
2. First towards third person: self-objectivation

Concerned Anthropologist

Remains outside of the tribe but is able to understand what is going on
(Garfinkel, 1969; Dennett, 1991)

Heterogening processes

1. A second person as a third one
2. A first person as the third one

Multiple Second Persons

Empathy
YOU-positions
Interpersonal

- Emotional compassion immediacy
- Cognitive analogizing mediation
- Ethical
- Gnoseological
- Coacher, spiritual friend
- Teacher scientific researcher

Community
WE-positions
Plurality of persons

- Emotional interactions
- Cognitive interactions
- Political (relations)
- Social (network)
- Coached community
- Community of co-researchers

2.5 *The Differentiated Plasticity of Second-Person Stances Proper: Taking into Account the Philosophy of Mind Approach*

Let us consider now the approach to the 'second person' in the philosophy of mind in order to see (1) how the phenomenological distinctions we just laid out are relevant to or can illuminate the way the issue of intersubjectivity is broached in the philosophy of mind; and (2) the extent to which the philosophy of mind approach is able to provide us with some other dimensions of the intersubjective experience that were not directly visible from a strict phenomenological point of view.

General presentation of the second-person issue in the framework of the philosophy of mind

In this context the problem is first stated as the problem of other minds. As such it is laden with a mentalistic presupposition and ruled by the theoretical postulate of a possible reflexive detachment which would amount to a communication between two Cartesian minds (Gallagher 2001, 83).

Two different approaches can be distinguished in this context: either we adopt a 'theory-theory' approach, according to which intersubjectivity is a matter of formulating a theory about the other person's mind, about his or her 'mind reality,' in order to explain his or her behaviour and subpersonal inferences (Leslie 1991); or we adopt a 'simulation-theory' approach, according to which we make 'as-if' belief and desire ascriptions by non-consciously simulating or rehearsing off-line what we would believe or desire and then ascribing it the other (Frith and Frith 1999; Goldman 1989; Gordon 1995). Simulation theory, it has been claimed (Gallese and Goldman 1998), is confirmed by the 'mirror-neuron' findings in neurophysiology.

But we can also make a radically pragmatic hypothesis that stands in sharp contrast with the initial mentalistic and theoretical postulate: then we understand others not as other minds but as other situated lived bodies. Such a practical and embodied primary intersubjectivity, as a key experience in understanding others in second-person interactions (Gallagher 2001), is attested to by the case of neonates (Meltzoff and Moore 1977) and has various features: (1) eye direction-detection and shared attention mechanisms (Baron-Cohen, 1995); (2) body-reading capabilities (Allison et al. 2000); and (3) affective coordination and emotional perceptions (Gopnick and Meltzhoff 1997).

We can thus present another philosophy of mind schema of the second-person issue (Figure 4).

[Figure 4: A philosophy of mind schema of the second-person issue]

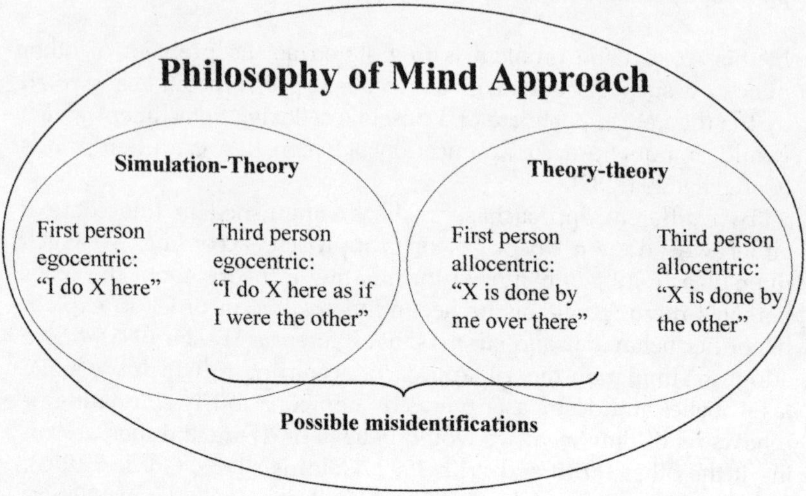

Philosophy of Mind Approach

Simulation-Theory

First person egocentric: "I do X here"

Third person egocentric: "I do X here as if I were the other"

Theory-theory

First person allocentric: "X is done by me over there"

Third person allocentric: "X is done by the other"

Possible misidentifications

Critical analysis

Whereas the extreme poles (first-person egocentric and third-person allocentric) are clearly identifiable and relatively univocal, the intermediate cases are more complex and problematic. From a phenomenological perspective they ought to be reversed: indeed the third-person egocentric is a process of self-objectivation of myself through the (imaginary/mental) presence of the other. So it is myself seen and treated from the outside of a third person. In that sense, it is objectifying and belongs to the third-person pole and only figuratively to the first-person one; on the contrary, the first-person allocentric is the apperception of an inner space within the first-person, which belongs to it but alters it. It brings about a dimension of self-alterity – or even of inner objectivity – within the I. Still, it remains a radically subjective sphere and therefore belongs to the first-person pole.

Moreover, what stands out in the philosophy-of-mind framework is the reducibility of the second-person problematic to a disguised first-person or third-person, as if no autonomous 'second' person could come out of such a framework. From an analytical point of view, indeed, no room is left for a genuine meaning of 'empathy' with its twofold (emotional and cognitive) components, either interpersonal or communitarian, and every relation and access to another subject/person is immediately seen in the mirror of myself or as a counter-object. Phenomenology in that respect attempts at unearthing an empathetic intersubjective space that is not reducible to first-person and third-person perspectives precisely because it encompasses them straightaway from the beginning.

2.6 The Intersubjective Phenomenology of the Second Persons at Work

Let us now show how these preliminary phenomenological distinctions between different crossed forms of empathy (emotional/cognitive on the one side, interpersonal/communitarian on the other side) are able to have a resonance with some empirical case studies, both inspiring them and being in their turn refined by them.

Limit-cases of empathy: Neonates, animals, psycho-pathology cases (schizophrenia, epilepsy) and foreigners

Empathy is not only a sharing of certain perceptual and embodied properties between the researcher and the subject, which would mean knowing you by knowing potential states of myself, and which would lead to a very narrow field of research, since it seems that we cannot relate therefore to people with different cognitive capabilities, for example psychotics, synesthetes, or even colour-blind people. Empathy has an extended meaning in Husserl himself, beyond the classical objection of empathy linked to resemblance and proximity. My empathetic experience with 'anormal' subjects (neonates, animals, the insane, or foreigners) is not only possible, it broadens my experience and affects me in a way that compels me to change my scope of experience and to change myself.[12]

---

12 For more on this issue, see Depraz (2001a).

Trained cases of empathy: Depth perception

Let us return to empathy in the concrete situation of experimental researcher and participant subject. In visual perception, some cases require a specific training, as in depth perception, where the percept is not a given but has to be accommodated through ocular convergence processes. In that respect the guidance can be external or internal, either in the case of another who plays the part of helping you focus your vision, or when you train yourself and increase your ability to become a guide for yourself. In both cases you learn to develop empathy, be it for your own self who is not accustomed to such a special vision, or be it the mentor who will empathize with you in order to increase in you such an ability at depth perception.[13]

Standard cases of empathy: Multistable perception

Even in what may appear as a standard visual experience, the part of apprenticeship is essential: after a time you do not see the phenomenon in the same manner (whether from trial to trial, or after repetition of the experience). You learn to notice aspects and dimensions that you were unable to grasp at first glance. A paradigmatic case is the well-known multistable perception experience. At first you are overwhelmed by the peculiar oscillation of the experience, but as you work along you can let go of the first impression (which does not mean it is irrelevant) and focus on the structure of the experience itself, not simply on the objects of perception. If the experimenter knows the experience himself,[14] and so can agree on (at least) some common aspects of it with his 'subjects,' he can play the role of a repository of possible dimensions of the experience,

13 For the setting and details of this experiment, see Lutz et al. (2002) and Lutz (2002).

14 Any literature search on such phenomena will reveal how studies that have described them rigorously begin with the researcher's observation of his own experience (Dutour 1760; Necker 1832; Breese 1909). The neurophenomenological program as initiated by Varela stresses a precise methodology to approach this critical issue of exploring experience itself (Varela 1996).

which the subject necessarily sees in a fragmented way. Simply by posing a more open vision of the experiment, the experimenter will thus help the subject to unfold aspects of the experience that he unavoidably could not see at first. On the other hand, the experimenter himself is necessarily exposed to *I–you* relations, which require that he adopt an open attitude towards his own experience. A concrete empathetic work of finding common categories and guiding the exploration of the experience is thus established. Training, both for the researcher and for the subject, far from being accessory, useless, or superficial, is therefore an essential component of what experience intrinsically is.

This last case exemplifies an important point we wish to highlight: here the guidance by the other seems to combine focused empathy with an open attitude. It is as if we had to do with a sort of synthesis of both forms of intersubjectivity into an *empathic openness*. As we will see, this attitude is in fact the guiding thread of an intersubjective approach to the understanding of human consciousness.

3. The Challenge of an Intersubjective Eidetics: The Practice of Empathic Openness in the Process of Categorizing

As we discussed above when constructing the relational schemes, actual experience is made of transitions between poles: we 'visit attitudes' or engage in points of view in a quite spontaneous way. 'Third persons' and 'first persons' are not closed but lean towards one another in an intersubjective ground. In such a region, we now wish to direct explicitly the actual process of categorizing into an empathic domain:

> We want to deal with the singularity of experience guided by the empathic resonance of common experience.

In addition to the traditional averaging procedures used as a way of establishing a valid category, we think it is pertinent to rely on the explicit intersubjective exploration of the experience itself.

3.1 *The Theoretical Core-Hypothesis*

As early as the Greek philosophers who initiated what is still today called philosophy, the necessity of identity and permanence has been considered a central prerequisite of science, insofar as science is supposed to rest on objectivity understood as both necessity and universality. Parmenides thus claimed the logic of identity as the only possibility for any stable objective knowledge, and Aristotle in his *Second Analytics* stressed that there is no science but of general objects. Movement and mobility therefore never could acquire the dignity of scientific objects: the deadlock of the Heraclitean flux as well as Zeno's paradoxes attest to this clearly. So the search for stability and invariants in empirical research is still telling about this ancient prerequisite for a permanent objectivity.

Against such a general agreement we would like to advance the reverse theoretical core hypothesis: in recent empirical research – but also tracing back to the discovery of quantum physics at the beginning of the century – instability has come to be seen as constitutive in dynamical systems (both theoretical and concrete) and is no longer considered a side-effect of a phenomenon initially defined by its stability (Nicolis and Prigogine 1989). In the same spirit, we would like to argue that the variability of the results given by different subjects of an experiment, or by the same subject at different moments through particular trials, does not necessarily mean that the framework of the experiment is inadequate or that the responses are 'subjective' and thus need to be eliminated as so much 'noise' (being considered as unintelligible and not trustable).[15] On the contrary, we want to take such an intra- and inter-variability seriously. Far from having to defend invariant objectivity against the so-called subjectivism of variability, we would like to show

15 We note that W. Köhler in the late twenties already stressed the importance of qualitative regularities as variations that need not be quantified in order to be valid: "But what about other instances in which either our problems are not of the quantitative type, or where we have no way to replace direct observation by observation of other facts which are better adapted to accurate measurement? Obviously, the various qualitative types of behavior are no less important than are the quantitative differences within a given type.... Again, while observing a

how a genuine claim for objectivity needs to take into account and to integrate the parameter of variability into its framework as an inherent component of it. This means that such a restyled objectivity is nourished by the very singularity and differences of the results of each subject, as much as by their standardization ('moyennage') through third-person protocols. Of course, we do not mean merely to reverse Aristotle's claim by asserting that there is no science but of variable singularity. Our leading hypothesis may be formulated as follows:

> Variability (which involves singularity) needs to be considered as a genuine component of scientific objectivity.

On the neurodynamic level it has begun to dawn on some scientists that singularities and thus variability are strict components of objectivity. Indeed, methodologies that seek to deal with the intrinsic variability of neural signals have been developed recently (Lachaux et al. 1999, 2002; Ionnides et al. 2002; David et al. 2002). Both at the level of signal analysis and brain imaging these approaches are concerned with treating variability as an actual non-trivial aspect of brain dynamics that cannot be simply disregarded by averaging trials across a given experimental session. Lachaux and collaborators developed a signal analysis method specifically adapted to the estimation of correlations between non-averaged signals. Both Ioannides and collaborators and David and collaborators present methods for estimating brain activity from magnetoencephalographic recordings (MEG) in single trial conditions. These last works explicitly discuss the issue of the singularity of experience and the importance in dealing with it experimentally.

On the phenomenological level, the description of the object as being first and foremost individuated in time leads to providing objectivity

---

man in a somewhat critical situation, it may be essential to observe whether he talks to us in a steady or a shaky voice. At the present time this is essentially a qualitative discrimination. In the future a method may be discovered by which the steadiness of voice can be measured. But, if such a method is to be properly used, we must still know from direct observation what we mean by steadiness or unsteadiness as temporary characteristics of voices. Otherwise, we should be in danger of measuring the wrong thing" (Köhler 1947, 38–39).

with an inherent touch of qualitative instant-singularity (Husserl 1954; Varela 1999).

Furthermore, it is interesting to note how the issue of the spontaneity of consciousness (Hanna and Thompson 2004) naturally finds strong resonance with our claim. It is evident that if we take seriously the spontaneous nature of consciousness, then no thorough understanding of human experience can be derived exclusively from averaged and therefore 'atemporal' observations.

3.2 A General Intersubjective Epistemological Methodology

If we want to be faithful to such a claim, we need to bring intersubjectivity in its fullest and broadest meaning into the scientific framework. What does this mean? In order to be clear, let us make a preliminary distinction between two main understandings of intersubjectivity: (1) in its narrow understanding (but also its strongest, more accurate, and thus more rigorous meaning), intersubjectivity is equivalent with empathy; (2) in its broad (but also more diffuse, evasive, and vague) meaning, it is synonymous with the ability to be open to the world and to others. In other word, whereas the first intersubjectivity is focal, reflexive, and formal as a method, the second one is more basic, spontaneous, and ecological. Why do we now claim a primacy for this latter meaning of intersubjectivity, whereas we first stressed empathy as a leading hypothesis for a second-person methodology? Because these two understandings of intersubjectivity are complementary, being situated at two different levels of our research program, and both need to be taken into account: (1) empathy is a well-structured and neatly layered experience, which nicely fits concrete and precisely defined empirical protocols; (2) the ability of openness, rather, meets the requirement of a more general epistemological methodology, which we would like now to discuss.

It may be useful, in this respect to rely on the holist methodology as it is remarkably accounted for by Kurt Goldstein in *The Organism*.[16]

16 "Simply put, we have endeavored to record, in an open-minded fashion, all phenomena. Pursuing this aim, there result three methodological postulates

The biologist provides us with a nice 'précis of methodology,' which is summarized in three main rules: (1) the experiencing subject needs to be open to all aspects of a given phenomenon: this implies an attitude without presuppositions, that is, a gesture of epoché of every prejudice. Consequently, one avoids taking one particular aspect into account simply because one does not know what to do with it or because it is astonishing or disturbing. The attitude in favour is on the contrary to become interested precisely in such an aspect, all the more meaningful because it first seems to be unintelligible. (2) Such an open attitude goes hand in hand with an attention to the multidimensionality of the experienced phenomenon, namely, of its very situatedness or individuation. What is here at stake is the component of the context/horizon (Gurwitsch 1957; Husserl, 1950a), or the environment/milieu (Husserl 1954; Von Uexküll 1956) as an intrinsic dimension of the phenomenon and not as something exterior to it. (3) Hence the final requisite of proceeding to a full categorizing description, which is attentive to and aware of the multifarious descriptive modalities and structural levels of the phenomenon. This categorizing indication is a direct consequence of the first two more experiential ones and is completely coherent with them.

---

equally valid for the examination of patients or animals. 1. Consider initially all the phenomena presented by the organism (in this case it may be a patient), giving no preference, in the description, to any special one. At this stage no symptom is to be considered of greater or less importance.... 2. The second methodological postulate concerns the correct description of the observable phenomena themselves. It was a frequent methodological error to accept what amounted to a mere description of the effect; but an effect might be ambiguous with respect to its underlying function. Therefore, only a thorough analysis of the causes of such effects, of success or failure in a given task, for example, can provide clarification.... Equally ambiguous are the negative results of a medical examination. The wrong response is too often judged to be a simple failure, whereas actually, under careful analysis, it may throw considerable light on the mental functions of the patient.... 3. The third methodological postulate we wish to stress is that no phenomenon should be considered without reference to the organism concerned and to the situation in which it appears" (Goldstein 1995, 37–40). See also Rosenthal and Visetti (1999).

The actual operation of this methodology is well illustrated in the above cases of depth perception and multistable perception, in which the attitude towards the experience follows this sequence of events and the relevant categories emerge from the open intersubjective exploration of the experience itself. We note that whereas in Goldstein's approach the experimenter does not appear as an explicit figure, the experimenter nonetheless cannot be overlooked, as it is he who exposes the subjects and mediates their description of the experience. In this sense it is he who is called in the first instance to adopt such an attitude towards experience. We would like now to insist on the third aspect of Goldstein's open methodology, while focusing on the process of production of the categories as they are emerging along with the experience we have of a given phenomenon.

3.3 *Taking the Genetic Process of Production of Categories into Account*

Usually we have an experience, we live it through, and then find words to account for it, including having difficulties in finding the right words to describe it. When we look at experience and description this way we consider them as two different and successive phases of the reflecting process. Husserl remains partly caught in such a view when he asserts in the *Cartesian Meditations*: "mute experience (*stumme Erfahrung*) still needs to be brought to the expression of its own meaning" (Husserl 1950b), whereas in some later manuscripts he speaks of a *denkend-schreibend* experience (Husserl 1973). Merleau-Ponty (1945, 1960) is less ambivalent when he insists on the intertwined dimensions of language and experience (Depraz 1999a). In short, we want to show that both the very experience as we first live and go through it, *and* the process of emergence of the categories that come to our mind in order to account for it, go hand in hand and coincide, are intertwined, and occur at the same moment. Or again, still more precisely: there is a common and continuous tissue between the immanent regularities of our experience and the first recognition of right categories.

Our attention is not necessarily directed toward such a categorial emergence at the very moment when we live the given experience on which we are focused. Hence the retrospective character of finding out categories of our experience in the necessary aftermath of the experience, whereas they were already there when it unfolded without us being able to notice them at the very moment when we were living it. Depending on the kind of training we have, we may become able to catch a glimpse of our categorizing activity while living a given experience. In any case, the point we want to make is that it is important to become attentive to the way our descriptive categories emerge to our mind as the experience unfolds and develops.

Thus the categories that enable us to describe our experience are given to us at different moments of the experience and along multifarious steps of their very constitution. In short, the categorizing process of my experience results as a gradual back-and-forth between certain categories and the unfolding of the experience. Hence the necessary *heuristic* character of such gradually and experientially produced categories, which correspond to different steps of categorization, before we are able to meet any validating categories, that is objectifying (universal and necessary) ones.

In this respect, we can make a preliminary distinction between at least three types of such heuristic experiential categories, all relative to the individual subject who is performing a given experience, and thus contributing to the very constitution of the categorizing process. Before any categories are given, there is their pre-givenness in a form of generative potentiality, which corresponds to the pre-conscious horizon of their experiential emergence: (1) expressive categories, which are both bodily, either strictly organic (sensory-kinesthetic) or behavioural (gestures and actions); (2) literary-linguistic categories, which belong to the *signifiant* dimension in Saussure's terminology, be they onomatopoeic words, images, and metaphors or again periphrases; 3) technical linguistic categories, which are culturally inherited in a broad sense, either as dialectical notions or educationally transmitted as a specific *Bildung* (domain of formation). The first two types of categories (expressive or literary-linguistic) are local, sensory-based, at the limit of communicability, while the third type may belong to an idiolectal minority or to an endogenous community, that is, are shared by a cultural group or a scientific community, as is illustrated below in the small table:

Table 1. Experiential Categories

DOMAINS	TYPES OF CATEGORIES			
	private	idiolectal	endogenous	universal
everyday	local	minority	community	world
scientific	a scientist alone	a cultural group	a scientific community	a scientific model
philosophical	sensation	intuition	ideologems	concepts

3.4 *Differentiated Criteria of Validating Variability*

In parallel with these different heuristic steps in the categorizing process, we have at least three distinct genetic criteria, which are helpful for increasing and enriching the objective validity of an experience:

(1) individual training as a means to develop more stability in the self-observation and self-description of one's own experience.
(2) intra-individual trials distributed along a specific temporal period in the life of the individual subject, which enable one to compare different performances of the same individual and better evaluate their more-or-less successful results.
(3) inter-individual sharing of a given experience, which allows the emergence of conflicting descriptions but also of convergent analysis.

3.4.1 Empirical case studies

In the two empirical case studies that we would like to detail now (both begun under the direction of the late Francisco Varela), these three genetic criteria are taken into account, but with a different emphasis.

Depth-perception (Lutz et al. 2002; Lutz 2002)

This experimental study (Lutz et al. 2002) aimed at giving an active role to the subject, whose neuronal dynamic was being registered at the very moment when he had to give a precise, first-hand descriptive account of his lived experience. The epistemological leading thread was the issue of how the subject's disciplined way of categorizing his experience is able to influence (modify, refine, renew) the endogenous empirical categories at work in the experimental protocol. The original thrust of this work was to focus on the different forms of individual training as multifarious forms of readiness for the emerging perceptual experience of depth (in a 2D autostereogram) and behavioural action (a button-press report). The authors distinguished among three main forms of subjective readiness that are shown to modulate perception and behaviour: (1) stable readiness, (2) fragmented readiness, and (3) unreadiness. Emphasis is put on the variability of the different trials over time, and the extent to which 'phenomenological clusters' corresponding to the three main forms of readiness can be used to categorize some of this variability.

Binocular rivalry (Cosmelli et al. 2002; David et al. 2002)

Here the experimental protocol stresses two linked aspects: (1) Each subject is also a sort of co-researcher, that is, can potentially guide the exploration of the experience. To do so, the subject first has to undergo the experience several times and describe it openly. The process of description is not done alone but is guided by open questions posed by the experimenter who plays the role of a midwife. The aim is to produce subjective invariants by undergoing the experience repeatedly and adopting at each moment the attitude discussed by Goldstein – stabilizing those aspects that seem *recurrent* throughout the experience, however simple or unimportant they may appear initially. (2) The group of subjects/co-researchers works within an empathic network in which the experimenter holds a particular status as a sort of central node on which all descriptions converge. Each subject gradually produces his or her own categories of description and the experimenter (a subject himself) has to leave aside his preconceptions about the experience and open towards those aspects that are invariant across subjects. The aimed-for result is the explicitly intersubjective production of categories during the development of the study.

3.5 *First Features of a Renewed Eidetics*

A first feature lies in the empathetic intervariability; a second in the genetic altering inner-variability; the third is the result of the first two and corresponds to the dynamics through which I am able to reach the structural 'variable invariants' of the experience. Such a dynamics proceeds in different steps: (1) sharing features beyond both variabilities; (2) producing contrasts *via* conflicts in variation; (3) stressing the strength of singularity as exemplary in contrast with standard quantified cases; (4) producing eidetic results as a mobile and plural cohesion of variable singularities.

On the basis of this renewed eidetics, we can distinguish among several different complementary modes of description: (1) expression; (2) account; (3) report; and (4) description. 'Expression' has quite a broad meaning: in the Merleau-Pontian understanding it designates all forms – linguistic, logical, but also bodily and gestural – through which experience is communicated; in the Wittgensteinian meaning, only the expressive enunciations correspond to the first-person perspective. 'Account,' according to the interactionist model promoted by Garfinkel (1967), is an individual first-person testimony as it is directed towards a second person. The relationship between two individual agents is given primacy, and the private world of the subject is put to the de-emphasized. 'Report' is meant to refer to the third-person verbalization of the experience of a given subject (Nisbett and Wilson 1977). 'Description' finally can be viewed in two different ways: either it phenomenologically corresponds to the generic term for the linguistic enunciation (formal and/or natural), be it first-person- or third-person-oriented, but with a stress on the inner truth of the first-person subject (according to the understanding of truth in phenomenology as a lived intuitive evidence); or it refers to the Wittgensteinian descriptive enunciations as so many trans-individual invariants, hence with a stress on the third-person perspective.

4. Conclusion: The Ontological Underpinning, a Trained Experiential Inter-Individuation

The notion of experience consequently presents three intertwined aspects: (1) Experience is always given as a singularity, as a unique event in space and time, in its very recurrence. It gives the time of the Husserlian 'living present' its whole strength. Such terms as 'pure experience' in a Jamesian context or 'bashô' in Nishida's philosophy refer to a similar experience of singularity in time. In a scientific framework, this aspect requires emphasizing the individuality of each experience. (2) Training is a constitutive component of the concept of experience as a singular experience. According to the general pragmatic spirit of our inquiry, we insist on the feasibility of experience being refinable, as opposed to any understanding of experience as limited a priori. (3) For experience to acquire its full sense and become an object of investigation it has to be shared. Far from diluting the singular character of individual experience, the fundamental attitude of empathic openness seeks to make it available in the search for common aspects. Training and inter-individuation are thus two complementary ways to gain a re-qualified objectivity of singularity.

It is in the context of experience so understood that we choose intersubjectivity as the departure point, the working ground. Our central point, given this working ground, is that empathy has in itself a methodological relevance for the understanding of human consciousness. It can be developed as such and can be enriched both theoretically and practically by two fundamental ideas: first, we are unavoidably historically coupled,[17] and second, we are constitutively open to ourselves and to the other. We think that the still intuitive experience of 'empathic resonance' is central and demands attention both on the operational level (within a given protocol that calls for it) and on the concrete experiential-experimental level (as an experience itself to be explored).

An intersubjective approach to experience puts forth an important methodological attitude. By adopting and exploring an explicit

---

17 As Thompson (2001) puts it: 'Between Ourselves.'

empathic constraint in the validation of experimental data, it requires a gesture of openness toward the other (and toward the question itself) from those who seek to meet on common ground. The gesture is not neutral, it offers explicit alternatives to ego-centred and mass-centred approaches. We are neither an isolated enclosed organism nor just a complex collective dynamic; we are unique individuals sharing a common world.

References

Allison, T., Q. Puce, and G. McCarthy (2000). "Social Perception from Visual Cues: Role of the STS region," *Trends in Cognitive Science* **4**: 267–78.

Baron-Cohen, S. (1995). *Mindblindness: An Essay on Autism and Theory of Mind*. Cambridge, MA: MIT Press.

Bitbol, M. (2002). "Science as if Situation Mattered," *Phenomenology and the Cognitive Sciences* **1**: 181–224.

Braddock G. (2001). "Beyond Reflection in Naturalized Phenomenology," *Journal of Consciousness Studies* **8(11)**: 3–16.

Breese, B. B. (1909). "Binocular Rivalry," *Psychological Review* **16**: 410–15.

Chalmers, D. J. (1995). "Facing Up to the Problem of Consciousness," *Journal of Consciousness Studies* **2(3)**: 200–19.

Cosmelli, D., O. David, J. P. Lachaux, L. Garnero, B. Renault, and F. J. Varela (2002). "Dynamic Neural Patterns Revealed by MEG/EEG during Visual Perception," *Clinical Neurophysiology* **113**, Supplement 1: Abstracts of the 11th European Congress of Clinical Neurophysiology, Barcelona, Spain, August 24–28, 2002.

David, O., L. Garnero, D. Cosmelli, and F. J. Varela (2002). "Estimation of Neural Dynamics from MEG/EEG Cortical Current Density Maps: Application to the Reconstruction of Large-Scale Cortical Synchrony," *IEEE Transactions Biomedical Engineering* **49**: 975–87.

Dennett, D. C. (1991). *Consciousness Explained*. Boston: Little Brown.

Dennett, D. C. (2002). "A Third-person Approach to Consciousness." Daewoo Lecture 1. Unpublished.

Depraz N. (1995). *Transcendance et incarnation. Le statut de l'intersubjectivité comme altérité à soi chez Husserl*. Paris: Vrin.

Depraz, N. (1999a). *Ecrire en phénoménologue*, Fougères, Encre Marine.

Depraz, N. (1999b). "The Phenomenological Reduction as Praxis." In F. J. Varela and J. Shear, eds., *The View from Within*, 95–110. Thorverton, UK: Imprint Academic.

Depraz N. (2001a). "The Husserlian Theory of Intersubjectivity as Alterology: Emergent Theories and Wisdom Traditions in the Light of Genetic Phenomenology." In E. Thompson, ed., *Between Ourselves: Second-Person Issues in the Study of Consciousness*, 169–78. Thorverton, UK: Imprint Academic. Also published in *Journal of Consciousness Studies* **8(5–7)**: 169–78.

Depraz, N. (2001b). "La phénoménologie, une pratique concrete," *Magazine Littéraire*, Novembre.

Depraz, N., and F. J. Varela (2003). "Au cœur du temps I," *Intellektika*, forthcoming.
Depraz, N., F. J. Varela, and P. Vermersch (2003). *On Becoming Aware: An Experiential Pragmatics*. Amsterdam: John Benjamins.
Dutour, E.-F. (1760). "Discussion d'une question d'optique," *Académie des Sciences: Mémoires de Mathématique et de Physique Présentes par Divers Savants* **3**: 514–30.
Frith, C. D., and U. Frith (1999). "Interacting Minds – A Biological Basis," *Science* **286**: 1692–95.
Gallagher, S. (2001). "The Practice of Mind: Theory, Simulation or Primary Interaction." In E. Thompson, ed., *Between Ourselves: Second-Person Issues in the Study of Consciousness*, 83–108. Thorverton, UK: Imprint Academic. Also published in *Journal of Consciousness Studies* **8(5–7)**: 83–108.
Gallagher, S., and J. Cole (2002). "Gesture Following Deafferentation: A Phenomenologically Informed Experimental Study." In: *Phenomenology and the Cognitive Sciences* **1**: 49–67.
Gallese, V., and A. I. Goldman (1998). "Mirror Neurons and the Simulation of Mind Reading," *Trends in Cognitive Science* **2**: 493–501.
Garfinkel, H. (1967). *Studies in Ethnomethodology*. Prentice Hall, NJ: Englewood Cliffs.
Gendlin, E. (1997). *Experiencing and the Creation of Meaning: A Philosophical and Psychological Approach to the Subjective*. Evanston, IL: Northwestern University Press.
Goldman, A. I. (1989). "Interpretation Psychologized," *Mind and Language* **4**: 161–85.
Goldstein, K. (1995). *The Organism*. New York: Zone Books.
Gopnik, A., and A. N. Meltzoff (1997). *Words, Thoughts, and Theories*. Cambridge, MA: MIT Press.
Gordon, R. M. (1995). "Simulation without Introspection or Inference from Me to You." In M. Davies and T. Stone, eds., *Mental Simulation: Evaluations and Applications*. Oxford: Blackwell.
Gurwitsch, A. (1957). *Théorie du champ de la conscience*. Paris: Desclée de Brouwer.
Hanna, R. and Thompson, E. (2004). "Neurophenomenology and the Spontaneity of Consciousness," in E. Thompson (ed.), *The Problem of Consciousness: New Essays in Phenomenological Philosophy of Mind. Canadian Journal of Philosophy*, Supplementary Volume.
Husserl, E. (1950a). *Ideen zu einer reinen Phänomenologie I*. Den Haag: M. Nijhoff.
Husserl, E. (1950b). *Cartesianische Meditationen, Hua I*. Den Haag: M. Nijhoff.

Husserl, E. (1954). *Die Krisis der europäischen Wissenschaften und die transzendentale Phänomenologie, Hua VI*. Den Haag: M. Nijhoff.
Husserl, E. (1973). *Zur Intersubjektivität (1905–1920), Hua XIII*. Den Haag: M. Nijhoff.
Ioannides, A. A., G. K. Kostopoulos, N. A. Lasarkis, L. Liu, T. Shibata, M. Schellens, V. Poghosyan, and A. Khurshudyan (2002). "Timing and Connectivity in the Human Somatosensory Cortex from Single Trial Mass Electrical Activity," *Human Brain Mapping* **15**: 231–46.
Jonas, H. (1966). *The Phenomenon of Life: Toward a Philosophical Biology*. Chicago: University of Chicago Press.
Köhler, W. (1947). *Gestalt Psychology*. New York: Liveright.
Latour, B. (1988). *Science in Action: How to Follow Scientists and Engineers through Society*. Cambridge, MA: Harvard University Press.
Lachaux, J. P., E. Rodriguez, J. Martinerie, and F. J. Varela (1999). "Measuring Phase Synchrony in Brain Signals," *Human Brain Mapping* **8(4)**: 194–208.
Lachaux, J. P., A. Lutz, D. Rudrauf, D. Cosmelli, M. Le Van Quyen, J. Martinerie, and F. J. Varela (2002). "Estimating the Time Course of Coherence between Single-Trial Brain Signals: An Introduction to Wavelet Coherence," *Clinical Neurophysiology* **32(3)**: 157–74.
Leslie, A. M. (1991). "The Theory of Mind Impairment in Autism: Evidence for a Modular Mechanism of Development." In A Whiten (ed.), *Natural Theories of Mind: Evolution, Development and Simulation of Everyday Mindreading*, 63–78. Oxford: Blackwell.
Levine, J. (1983). "Materialism and Qualia: The Explanatory Gap." *Pacific Philosophical Quarterly* **64**: 354–61.
Lutz, A. (2002). "Toward a Neurophenomenology as an Account of Generative Passages: A First Empirical Case Study," *Phenomenology and the Cognitive Sciences* **1**: 133–67.
Lutz, A., J.-P. Lachaux, J. Martinerie, and F. J. Varela (2002). "Guiding the Study of Brain Dynamics Using First-Person Data: Synchrony Patterns Correlate with Ongoing Conscious States during a Simple Visual Task," *Proceedings of the National Academy of Sciences USA* **99**: 1586–91.
Meltzhoff, A., and M. K. Moore (1977). "Imitation of Facial and Manual Gestures by Human Neonates," *Science* **198**: 83–99.
Merleau-Ponty, M. (1945). *Phénoménologie de la perception*, Paris: Gallimard.
Merleau-Ponty, M. (1960). *Signes*. Paris: Gallimard.
Nagel, T. (1974). "What is it like to be a bat?" *Philosophical Review* **83**: 435–50.
Nam-in-Lee (1998). "Edmund Husserl's Phenomenology of Mood." In N. Depraz and D. Zahavi, eds., *Alterity and Facticity. New Perspectives on Husserl*. Dordrecht: Kluwer.

Necker, L. A. (1832). "Observations on some remarkable phaenomena seen in Switzerland; and on an optical phaenomenon which occurs on viewing a figure of a crystal or geometrical solid," *London and Edinburgh Philosophical Magazine and Journal of Science* **1**: 329–37.

Nicolis, G., and I. Prigogine (1989). *Exploring Complexity: An Introduction*. New York: W. H. Freeman and Co.

Nisbett, R., and T. D. Wilson (1977). "Telling More than We Can Know: Verbal Reports on Mental Processes," *Psychological Review* **84(3)**: 231–59.

Petitmengin, C. (2001). *L'expérience intuitive*. Paris: L'harmattan.

Rosenthal, V., and Y.-M. Visetti (1999). "Sens et temps de la Gestalt," *Intellectica* **1**: 147–227.

Scheler, M. (1970). *The Nature of Sympathy*. Trans. P. Heath. Hamden, CO: Archon Books.

Schutz, A. (1967) *The Phenomenology of the Social World*. Trans. G. Walsh and F. Lehnert. Evanston, IL: Northwestern University Press.

Spiegelberg, H. (1971). *Doing Phenomenology. Essays on and in Phenomenology*. The Hague: M. Nijhoff.

Spiegelberg, H. (1995). "Towards a Phenomenology of Imaginative Understanding of Others," *Proceedings of the 11th International Congress of Philosophy*, Brussels, 1995, IV, 232–39.

Stengers, I. (1994). *L'invention des sciences modernes*. Paris: La Découverte.

Stumpf, C. (1883). *Tonpsychologie*, Leipzig: R. Hirzel, 1883.

Thompson, E., ed. (2001). *Between Ourselves: Second-Person Issues in the Study of Consciousness*. Thorverton, UK: Imprint Academic.

Varela, F. J., (1979). *Principles of Biological Autonomy*, New York: Elsevier North Holland.

Varela, F. J. (1991). "Organism: A Meshwork of Selfless Selves." In A. Tauber, ed., *Organism and the Origin of Self*, 79–107. Dordrecht: Kluwer.

Varela, F. J., (1996). "Neurophenomenology: A Methodological Remedy for the Hard Problem," *Journal of Consciousness Studies* **3(4)**: 330–49.

Varela, F. J., (1997). "The Naturalization of Phenomenology as the Transcendence of Nature," *Alter* **5**: 355–81.

Varela, F. J. (1999). *Ethical Know-how*, Stanford: Stanford University Press.

Varela, F. J., and J. Shear, eds. (1999). *The View from Within*. Thorverton, UK: Imprint Academic.

Varela, F. J., (2000). "The Specious Present: A Neuro-Phenomenology of Time-Consciousness." In J. Petitot, F. Varela, J.-M. Roy, and B. Pachoud, eds., *Naturalizing Phenomenology*. Stanford: Stanford University Press.

Varela, F. J., and N. Depraz (2000). "At the Source of Time: Valence and the Constitutional Dynamics of Affect," *Arobase*. Electronic Journal: http://www.arobase.to

Varela, F. J., and N. Depraz (2003). "Imagining: Embodiment, Phenomenology and Transformation." In A. Wallace, ed., *Breaking the Ground: Essays on Tibetan Buddhism and the Natural Sciences*. New York: Columbia University Press.

Von Uexküll, J. (1956). *Streifzüge durch die Umwelten von Tieren und Menschen – Bedeutungslehre*, Hamburg: Rowohlt Verlag.

Schütz, A. (1962–66). *Collected Papers*. The Hague: M. Nijhoff.

Zahavi, D. (1996). *Husserl und die transzendentale Intersubjektivitat. Eine Antwort auf die sprachpragmatische Kritik*. Dordrecht/Boston/London: Kluwer.

Zahavi, D. (2001). "Beyond Empathy: Phenomenological Approaches to Intersubjectivity." In E. Thompson, ed., *Between Ourselves: Second-Person Issues in the Study of Consciousness*. Thorverton, UK: Imprint Academic, 151–67. Published simultaneously as *Journal of Consciousness Studies* **8(5–7)**: 151–67.

Notes on Contributors

DIEGO COSMELLI is a native of Santiago, Chile. He studied biochemistry at the Universidad de Chile, obtaining his degree in 1998. He begun his doctoral work in cognitive science at the Ecole Polytechnique in 2000 under the direction of the late Francisco Varela, and received his Ph.D. in 2004. His main interests centre on the relationship between brain dynamics and human experience.

NATALIE DEPRAZ earned her Ph.D. in Philosophy in 1993, focusing on the writings of Husserl and the phenomenology of intersubjectivity. Since 1997 she has been directrice de programme au Collège International de Philosophie (Paris), and since 2000 she has been maître de conférences en philosophie à l'Université de la Sorbonne (Paris IV). Her publications include *Transcendence et incarnation. Le statut de l'intersubjectivité comme altérité à soi chez Husserl* (Paris: Vrin, 1995), *Alterity and Facticity: New Perspectives on Husserl*, edited with Dan Zahavi (Dordrecht: Kluwer, 1998), *Lucidité du corps. De l'empiricisme transcendental en phénoménologie* (Dordrecht: Kluwer, 2001), and *On Becoming Aware: An Experiential Pragmatics* (with Francisco Varela and Pierre Vermersch) (Amsterdam: John Benjamins, 2003). She is the journal editor of *Alter, revue de phénomenologie* and *Phenomenology and the Cognitive Sciences* (with Shaun Gallagher).

DENIS FISETTE is Professor of Philosophy at the University of Québec at Montréal. He is the author (with Pierre Poirier) of *Philosophie de l'esprit: État des lieux* (Paris: Vrin, 2000), and editor of *Consciousness and Intentionality: Models and Modalities of Attribution* (Dordrecht: Kluwer, 1999).

SHAUN GALLAGHER is Professor and Chair of the Philosophy Department and Director of the Cognitive Science Program at the University of Central Florida. He is co-editor of the interdisciplinary journal *Phenomenology and the Cognitive Sciences*. His research and teaching interests include phenomenology and philosophy of mind, cognitive science, hermeneutics (theory of interpretation), and theories of the self and personal identity. His new book, *How the Body Shapes the Mind*, will be published by Oxford University Press in 2004. He previously authored two books: *Hermeneutics and Education* (1992) and *The Inordinance of Time* (1998). Dr. Gallagher has edited and coedited several volumes, including *Models of the Self* with Jonathan Shear (1999) and *Hegel, History, and Interpretation* (1997). He has published in various professional journals and anthologies on topics such as the embodied self, social cognition, and psychopathology.

ROBERT HANNA is Associate Professor of Philosophy at the University of Colorado, Boulder. He received his Ph.D. in philosophy from Yale in 1989, and his honours B.A. in philosophy from the University of Toronto. He is a Canadian by birth and a naturalized U.S. citizen. His areas of research and specialization include (1) the history of modern philosophy (especially Kant, the analytic tradition, and the phenomenological tradition), (2) the philosophy of mind and cognition (especially consciousness, mental causation, and rationality), and (3) ethics (especially the overlap between ethics and the philosophy of mind). In connection with those areas, he is the author of *Kant and the Foundations of Analytic Philosophy* (Oxford: Clarendon Press, 2001) and a number of journal articles.

JEAN-MICHEL ROY is currently teaching philosophy at the Ecole Normale Supérieure de Lettres et Sciences Humaines in Lyon. An affiliate of the Paris Husserl Archives (CNRS) and of CREA (CNRS), his work is centred on the historical analysis of the analytical turn of philosophy, as well as on the foundations of cognitive science. In both areas, he seeks to re-evaluate the role of Husserlian phenomenology. His recent publications include *Naturalizing Phenomenology: Issues on Contemporary Phenomenology and Cognitive Science* (Stanford University Press, 1999).

EVAN THOMPSON is Associate Professor in the Department of Philosophy at York University, where he holds a Canada Research Chair in Cognitive Science and the Embodied Mind. He has published numerous articles in philosophy of mind and the theoretical foundations of cognitive science. He is a co-author (with Francisco J. Varela and Eleanor Rosch) of *The Embodied Mind: Cognitive Science and Human Experience* (MIT Press, 1991) and author of *Colour Vision: A Study in Cognitive Science and the Philosophy of Perception* (Routledge Press 1995). In addition to the present volume, he has edited two collections, *Between Ourselves: Second-Person Issues in the Study of Consciousness* (Imprint Academic, 2001) and (with Alva Noë) *Vision and Mind: Selected Readings in the Philosophy of Perception*. He is currently completing a book entitled *Radical Embodiment: The Lived Body in Biology, Human Experience, and the Sciences of Mind* (to be published by Harvard University Press).

FRANCISCO VARELA (1946–2001) was director of research at the Centre National de Recherche Scientifique (LENA and CREA, Paris). A neurobiologist, he published widely in cognitive neuroscience, theoretical biology, immunology, and the theoretical foundations of cognitive science. His books include *Principles of Biological Autonomy* (Elsevier 1979), *Autopoiesis and Cognition: The Realization of the Living* (with Humberto Maturana), Boston Studies in the Philosophy of Science, v. 42 (D. Reidel, 1980), and *The Embodied Mind: Cognitive Science and Human Experience* (with Evan Thompson and Eleanor Rosch) (MIT Press, 1991).

DAN ZAHAVI is Professor at and Director of the Danish National Research Foundation: Center for Subjectivity Research, at the University of Copenhagen. He has published widely in phenomenology and philosophy of mind and is the author and editor of fourteen books, including *Husserl und die transzendentale Intersubjektivität* (Kluwer 1996), *Self-awareness and Alterity* (Northwestern 1999), *Exploring the Self* (John Benjamins 2000), *One Hundred Years of Phenomenology* (Kluwer 2002), and *Husserl's Phenomenology* (Stanford 2003).

Index